Keine original LEGO® Bauanleitungen. LEGO® hat dieses Buch weder gesponsert noch autorisiert. In diesem Buch enthaltene Bauanleitungen beruhen nicht auf einer Lizenz oder Billigung durch die Rechtsinhaber der LEGO®-Kennzeichen und LEGO®-Produkte. Die Benutzung der LEGO®-Marken und -Produkte erfolgt ausschließlich zur Illustration der in diesem Buch enthaltenen Bauideen. Sämtliche Rechte an Gestaltungen und Marken stehen allein dem jeweiligen Rechtsinhaber zu.

Never to old to play with bricks...

Martin Ludwig & Frank Müller

Miniscale Wehrmacht vehicles instructions

to be build out of LEGO® Bricks

www.tredition.de

© 2016 Martin Ludwig & Frank Müller
Umschlag, Illustration: Frank Müller

Verlag tredition GmbH, Hamburg

ISBN
Paperback: 978-3-7323-7917-0
Hardcover: 978-3-7323-7918-7
e-Book: 978-3-7323-7919-4

Printed in Germany

Das Werk, einschließlich seiner Teile, ist urheberrechtlich geschützt. Jede Verwertung ist ohne Zustimmung des Verlages und des Autors unzulässig. Dies gilt insbesondere für die elektronische oder sonstige Vervielfältigung, Übersetzung, Verbreitung und öffentliche Zugänglichmachung.

Keine original LEGO® Bauanleitungen. LEGO® hat dieses Buch weder gesponsert noch autorisiert. In diesem Buch enthaltene Bauanleitungen beruhen nicht auf einer Lizenz oder Billigung durch die Rechtsinhaber der LEGO®-Kennzeichen und LEGO®-Produkte. Die Benutzung der LEGO®-Marken und -Produkte erfolgt ausschließlich zur Illustration der in diesem Buch enthaltenen Bauideen. Sämtliche Rechte an Gestaltungen und Marken stehen allein dem jeweiligen Rechtsinhaber zu.

Alle hier abgebildeten Modelle sind bei der DPMA registriert.

Ebenfalls im tredition Verlag erschienen:
Bauanleitung Tiger & Königstiger

Also published by tredition:
Building instruction Tiger & Kingtiger tank

ISBN
Paperback: 978-3-7323-1027-2
Hardcover: 978-3-7323-1028-9
e-Book: 978-3-7323-1029-6

Ebenfalls im tredition Verlag erschienen:
WW2 Wehrmacht custom building instructions

Also published by tredition:
WW2 Wehrmacht custom building instructions

ISBN
Paperback: 978-3-7323-4183-2
Hardcover: 978-3-7323-4184-9
e-Book: 978-3-7323-4185-6

Martin Ludwig und Frank Müller sind zwei begeisterte Hobby AFOLs (Adult Fan of LEGO®) die sich auf den Nachbau von Militär Fahrzeugen spezialisiert haben. Die Fahrzeuge sind passend für die Größe der Figuren ausgelegt.

Alles begann mit einer Idee. Nach erfolgreichem Verkauf einiger Fahrzeuge bei eBay® wurde die Produktpalette stetig erweitert. Je nach Größe dauert das Design eines Fahrzeuges einige Wochen. Hier achten wir zum größten Teil auf die Funktionalität und die Verfügbarkeit der Teile. Natürlich können wir alles noch besser und komplexer bauen, aber nicht jedem Kunden stehen unsere Möglichkeiten zur Verfügung. Nach der Veröffentlichung der ersten beiden Baubücher liegt der Fokus dieser Bauanleitungen auf Minicale Fahrzeugen. In diesem Buch enthalten sind die Bauanleitungen für:

- Flak 88 Mini (Seite 8 - 18)
- Kingtiger Mini (Seite 19 - 37)
- Panzer IV Mini (Seite 38 - 50)
- SDKFZ 9 Mini (Seite 51 - 60)
- Stug Mini (Seite 61 - 69)
- Tiger Mini (Seite 70 - 85)

Martin Ludwig and Frank Müller are two hobby LEGO® builders from Germany. They spend their time building military models out of LEGO® bricks. It all started with an idea. After selling a few vehicles on eBay®, they started a successful business. The design of each vehicle takes a few weeks. Functionality and accurate size of vehicles are the main goals. After the successful release of their first books, they have now finished their third instruction book focused on mini vehicles. This book features:

- Anti Aircraft Gun 88 Mini (Page 8 - 18)
- Kingtiger Mini (Page 19 - 37)
- Panzer IV Mini (Page 38 - 50)
- SDKFZ 9 Mini (Page 51 - 60)
- Stug Mini (Page 61 - 69)
- Tiger Mini (Page 70 - 85)

Flak 88 Mini

Zum Bau des Modells benötigen Sie ca. 70 LEGO® Bausteine.
Länge: ca. 13,7 cm Breite: ca. 7,9 cm Höhe: ca. 5,1 cm

Anti Aircraft Gun 88 Mini

Requires approx. 70 LEGO® bricks.
length: ca. 13.7 cm width: ca. 7.9 cm height: ca. 5.1 cm

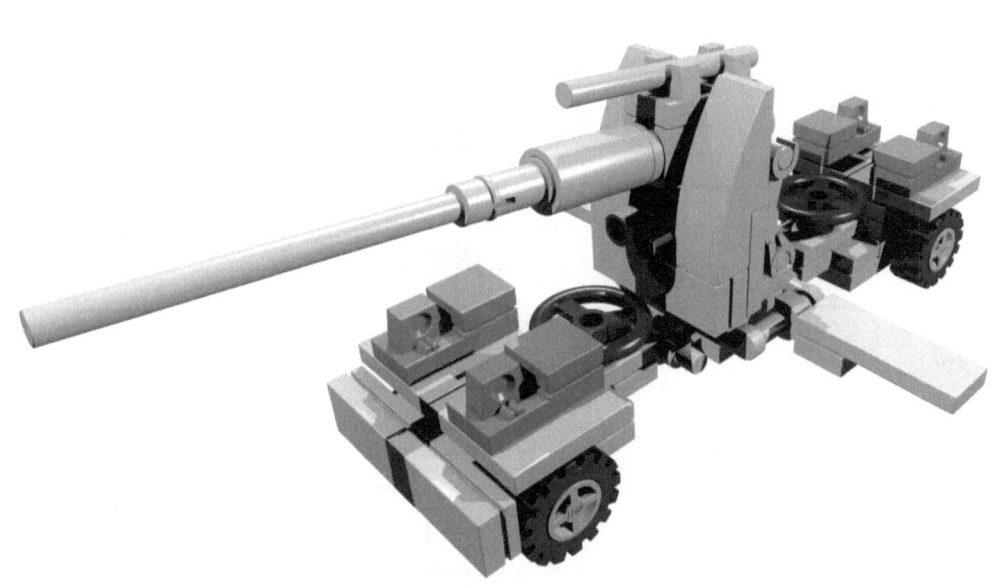

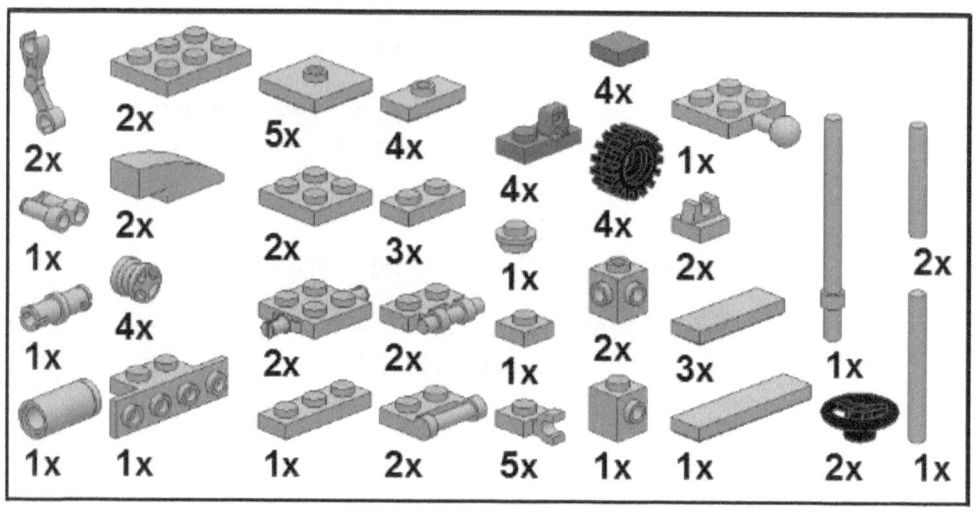

BLink ID	Color	Qty	Description
3021	Grey	2	Plate 2 x 3
4600	Grey	2	Plate, Modified 2 x 2 with Wheels Holder
3022	Grey	2	Plate 2 x 2
3023	Grey	3	Plate 1 x 2
2436	Grey	1	Bracket 1 x 2 - 1 x 4
2431	Grey	1	Tile 1 x 4
3794	Grey	4	Plate, Modified 1 x 2 with 1 Stud (Jumper)
87580	Grey	5	Plate, Modified 2 x 2 with Groove and 1 Stud in Center
4624	Grey	4	Wheel 8mm D. x 6mm
3641	Black	4	Tire Offset Tread Small
30383	Dark Grey	4	Hinge Plate 1 x 2 Locking with 1 Finger On Top
3070b	Dark Grey	4	Tile 1 x 1 with Groove
3731	Grey	1	Plate, Modified 2 x 2 with Towball
2540	Grey	2	Plate, Modified 1 x 2 with Handle on Side
48336	Grey	2	Plate, Modified 1 x 2 with Handle on Side
4733	Grey	2	Brick, Modified 1 x 1 with Studs on 4 Sides
6019	Grey	5	Plate, Modified 1 x 1 with Clip Horizontal
30663	Black	2	Vehicle, Steering Wheel Small, 2 Studs Diameter
63864	Grey	3	Tile 1 x 3

BLink ID	Color	Qty	Description
87994	Grey	2	Bar 3L
30377	Grey	2	Arm Mechanical, Battle Droid
4073	Grey	1	Plate, Round 1 x 1 Straight Side
87087	Grey	1	Brick, Modified 1 x 1 with Stud on 1 Side
2555	Grey	2	Tile, Modified 1 x 1 with Clip
30374	Grey	1	Bar 4L (Lightsaber Blade / Wand)
3673	Grey	1	Technic, Pin without Friction Ridges Lengthwise
63965	Grey	1	Bar 6L with Stop Ring
75535	Grey	1	Technic, Pin Connector Round (Pin Joiner Round)
30162	Grey	1	Minifig, Utensil Binoculars Town
50950	Grey	2	Slope, Curved 3 x 1 No Studs
3623	Grey	1	Plate 1 x 3
3024	Grey	1	Plate 1 x 1

1

2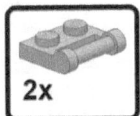

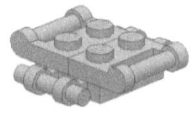

www.ww2custombrickmodels.de

3

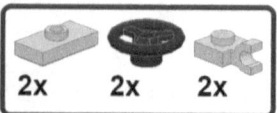

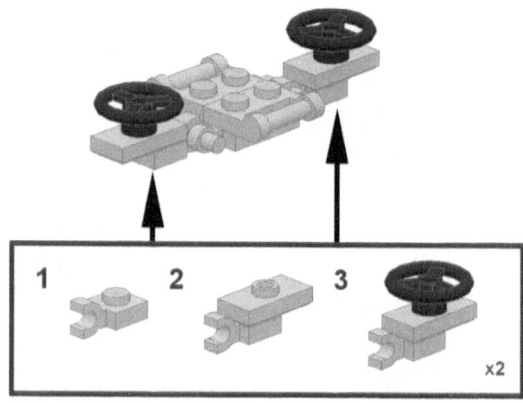

4

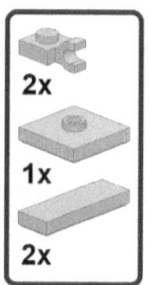

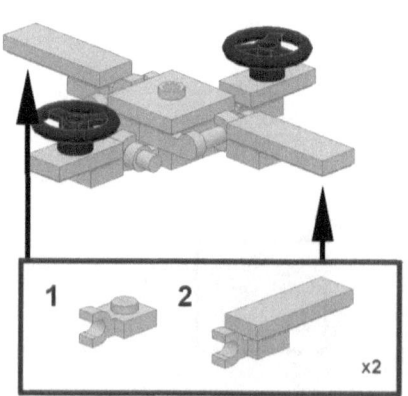

5

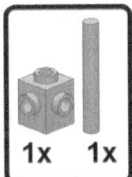

6

7

www.ww2custombrickmodels.de

8

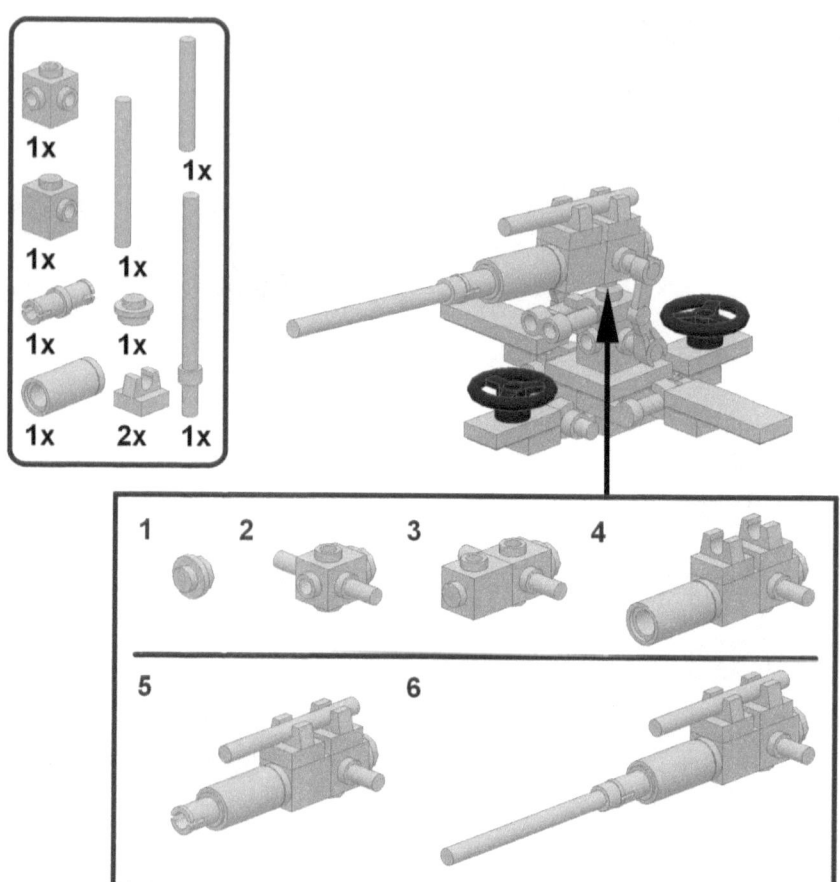

9

1x
2x
2x
1x
1x

1 2 3

1

1x

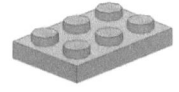

www.ww2custombrickmodels.de 13

2
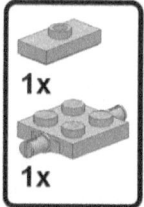

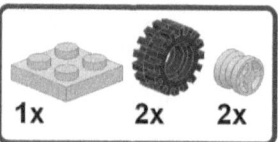

3

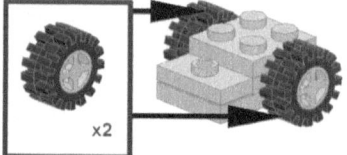

4
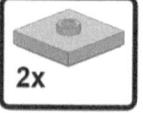

5

6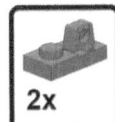

7

10

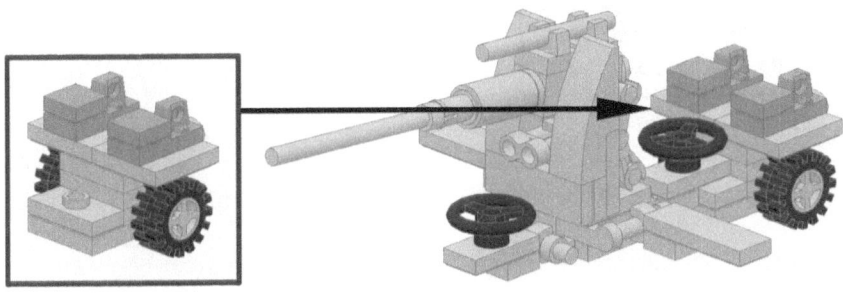

www.ww2custombrickmodels.de 15

1

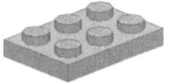

2

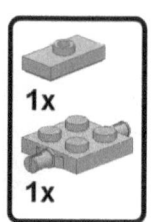

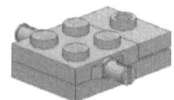

3

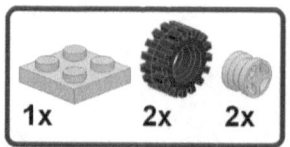

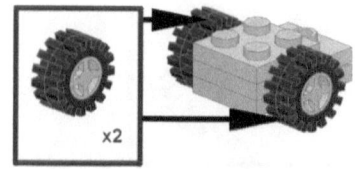

4

5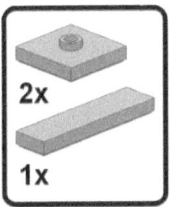

6

www.ww2custombrickmodels.de

7

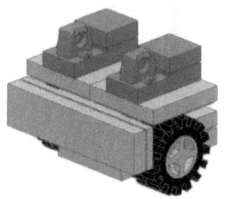

11

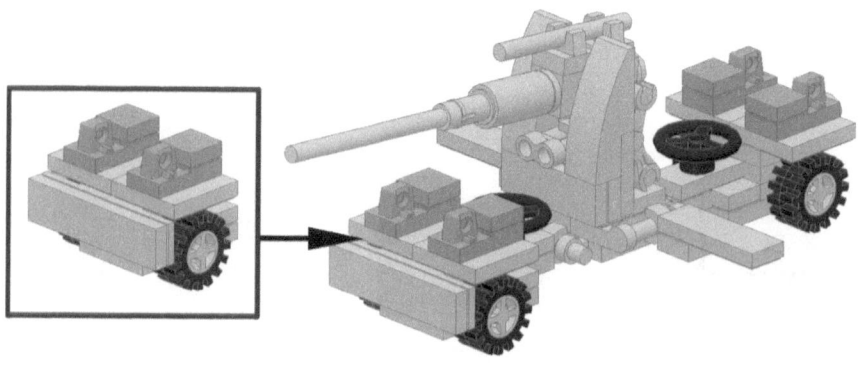

Kingtiger Mini

Zum Bau des Modells benötigen Sie ca. 233 LEGO® Bausteine.
Länge: ca. 13,0 cm Breite: ca. 5,1 cm Höhe: ca. 4,7 cm

Kingtiger Mini

Requires approx. 233 LEGO® bricks.
length: ca. 13.0 cm width: ca. 5.1 cm height: ca. 4.7 cm

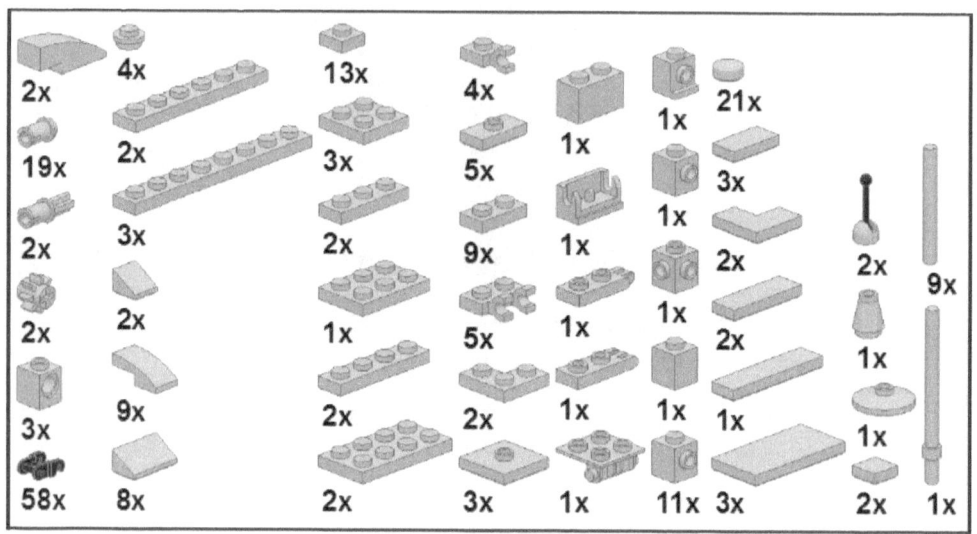

BLink ID	Color	Qty	Description
98138	Grey	21	Tile, Round 1 x 1
3647	Grey	2	Technic, Gear 8 Tooth Type 1
3749	Grey	2	Technic, Axle Pin without Friction Ridges Lengthwise
87079	Grey	3	Tile 2 x 4
3020	Grey	2	Plate 2 x 4
6541	Grey	3	Technic, Brick 1 x 1 with Hole
3710	Grey	2	Plate 1 x 4
3024	Grey	13	Plate 1 x 1
4276	Grey	1	Hinge Plate 1 x 2 with 2 Fingers
87087	Grey	11	Brick, Modified 1 x 1 with Stud on 1 Side
2431	Grey	1	Tile 1 x 4
4274	Grey	19	Technic, Pin 1/2
50950	Grey	2	Slope, Curved 3 x 1 No Studs
11477	Grey	9	Slope, Curved 2 x 1 No Studs
3022	Grey	3	Plate 2 x 2
4733	Grey	1	Brick, Modified 1 x 1 with Studs on 4 Sides
3023	Grey	9	Plate 1 x 2

BLink ID	Color	Qty	Description
3004	Grey	1	Brick 1 x 2
3794	Grey	5	Plate, Modified 1 x 2 with 1 Stud (Jumper)
4740	Grey	1	Dish 2 x 2 Inverted (Radar)
4589	Grey	1	Cone 1 x 1 without Top Groove
47905	Grey	1	Brick, Modified 1 x 1 with Studs on 2 Sides
63965	Grey	1	Bar 6L with Stop Ring
2420	Grey	2	Plate 2 x 2 Corner
54200	Grey	2	Slope 30 1 x 1 x 2/3
3623	Grey	2	Plate 1 x 3
3460	Grey	3	Plate 1 x 8
3021	Grey	1	Plate 2 x 3
4275	Grey	1	Hinge Plate 1 x 2 with 3 Fingers
63864	Grey	2	Tile 1 x 3
3069b	Grey	3	Tile 1 x 2 with Groove
87580	Grey	3	Plate, Modified 2 x 2 with Groove and 1 Stud in Center
3666	Grey	2	Plate 1 x 6
3070b	Grey	2	Tile 1 x 1 with Groove
85984	Grey	8	Slope 30 1 x 2 x 2/3
4073	Grey	4	Plate, Round 1 x 1 Straight Side
14719	Grey	2	Tile 2 x 2 Corner
3711	Black	58	Technic, Link Chain
4592c02	Grey	2	Lever Small Base with Black Lever
6134	Grey	1	Hinge Brick 2 x 2 Top Plate Thin
4070	Grey	1	Brick, Modified 1 x 1 with Headlight
60470	Grey	5	Plate, Modified 1 x 2 with Clips Horizontal
3937	Grey	1	Hinge Brick 1 x 2 Base
6019	Grey	4	Plate, Modified 1 x 1 with Clip Horizontal
3005	Grey	1	Brick 1 x 1
30374	Grey	9	Bar 4L (Lightsaber Blade / Wand)

1

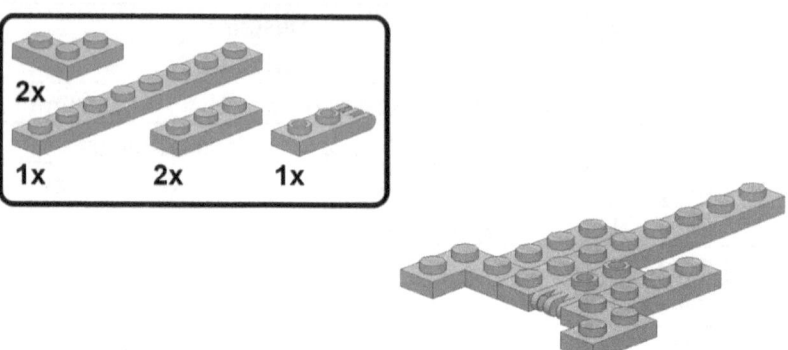

2

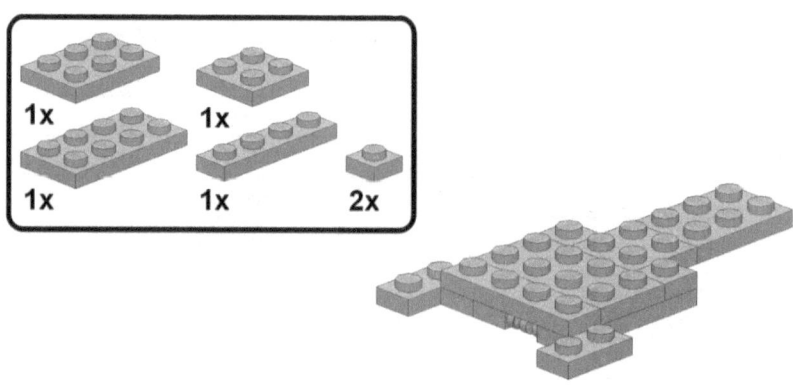

3

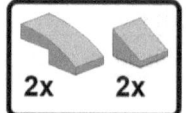

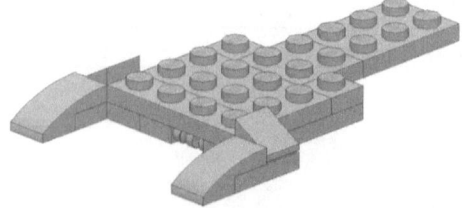

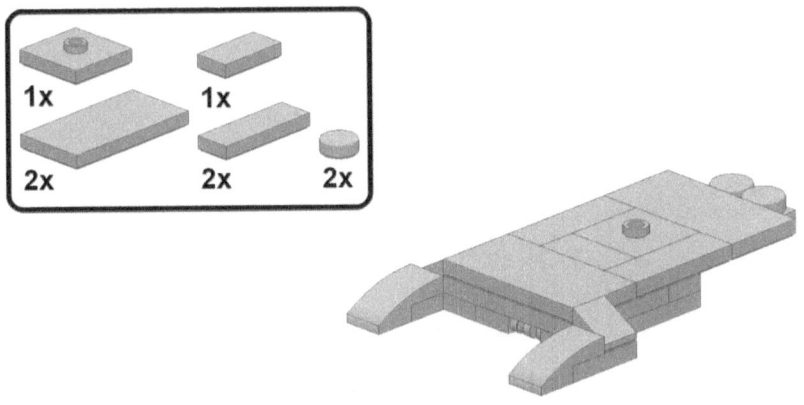

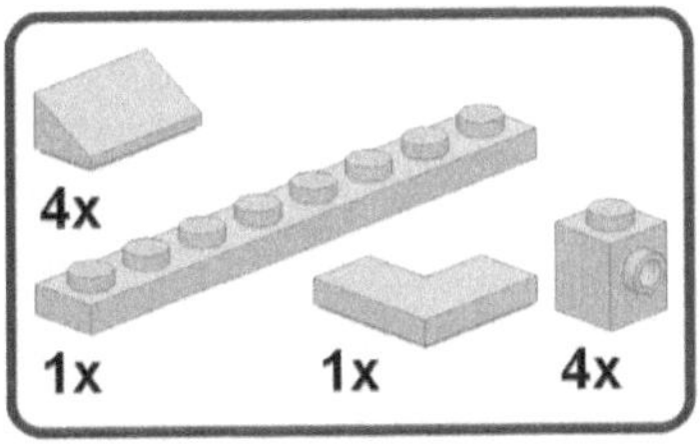

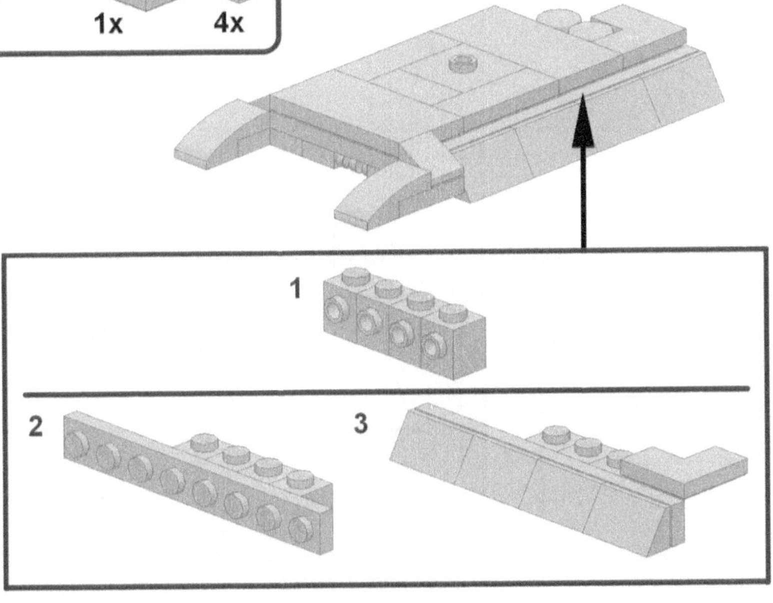

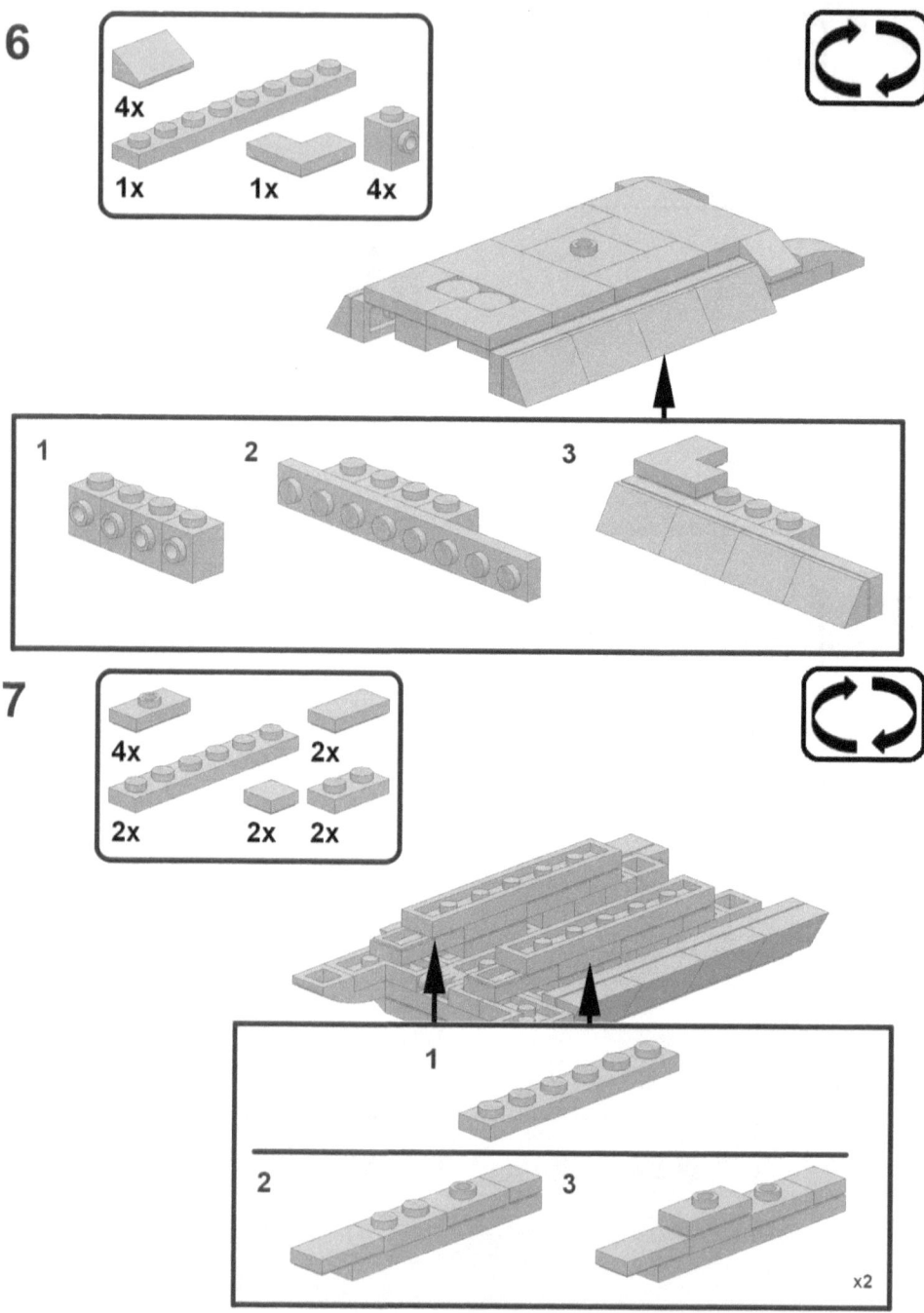

8

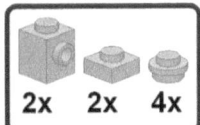

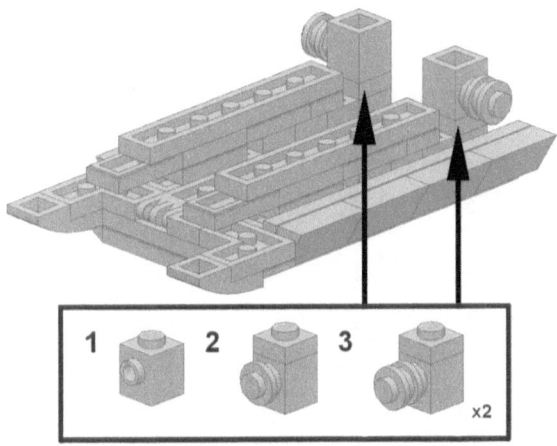

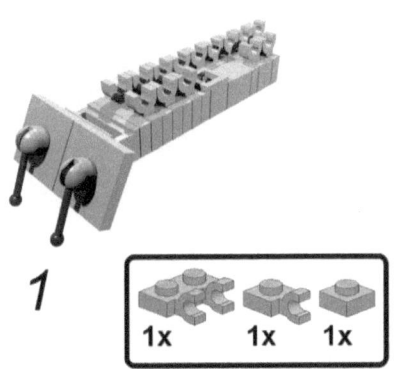

1

www.ww2custombrickmodels.de

2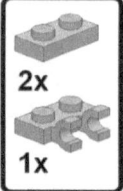

3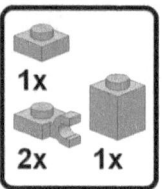

4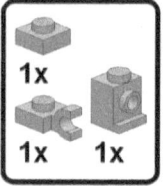

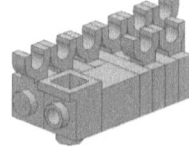

5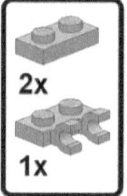

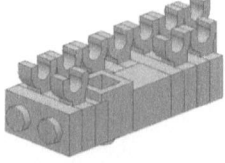

www.ww2custombrickmodels.de

6

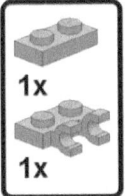

7

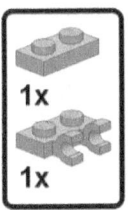

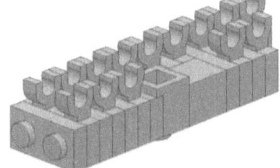

8

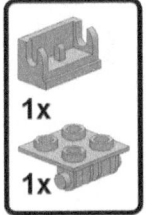

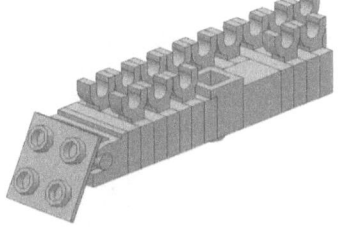

9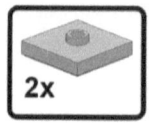

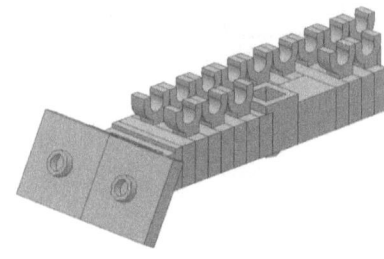

10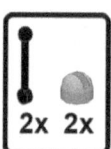

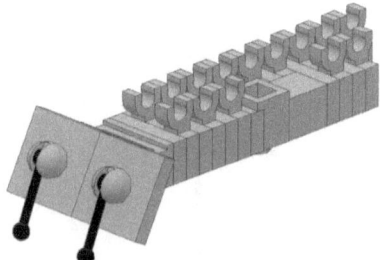

9

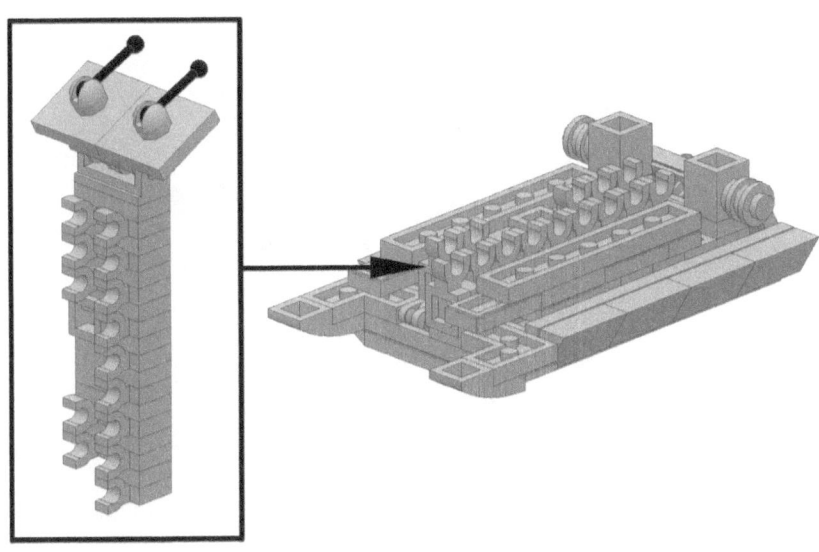

10

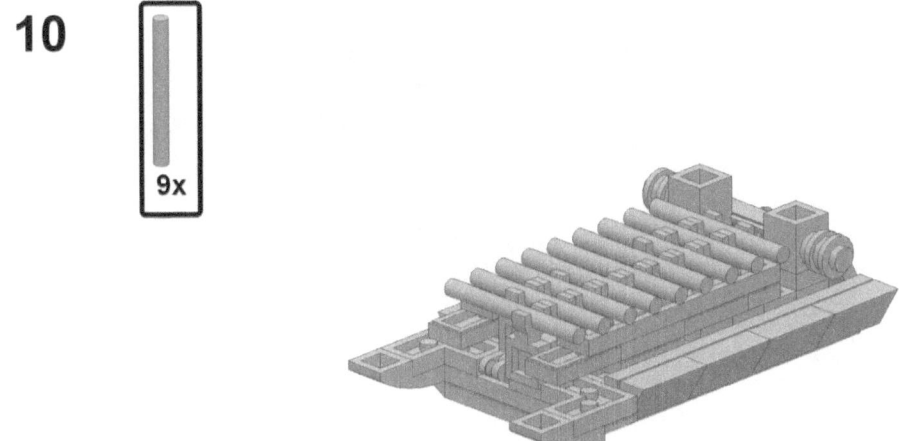

9x

11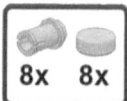

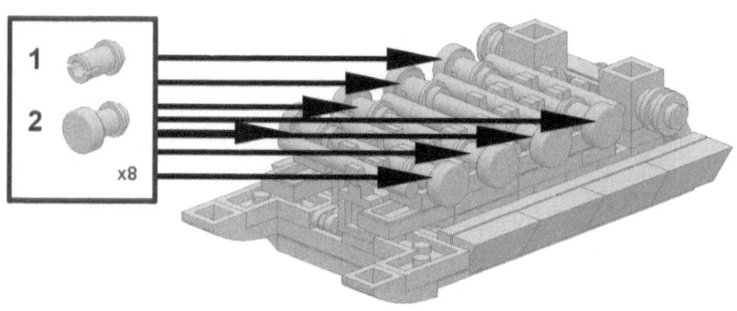

12

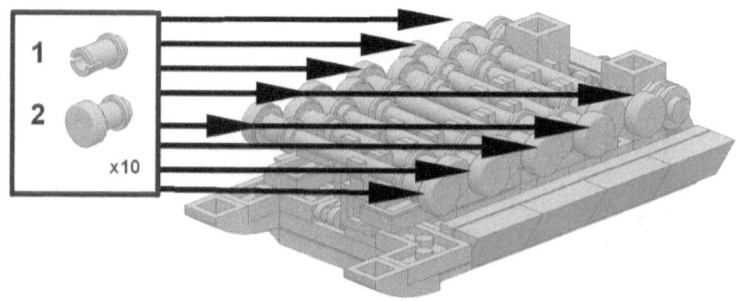

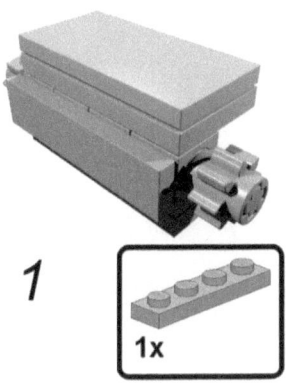

1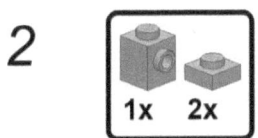

2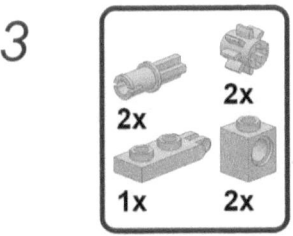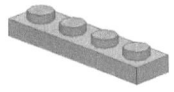

3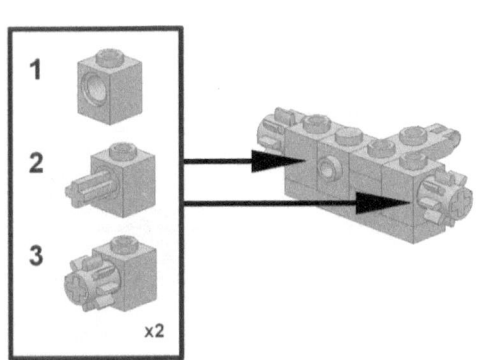

www.ww2custombrickmodels.de

4

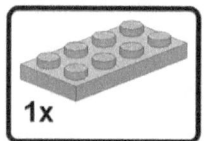

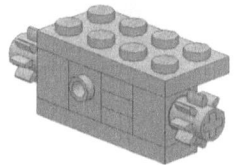

5

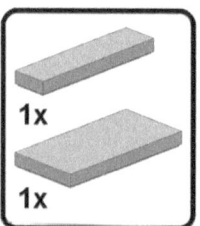

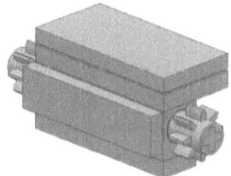

13

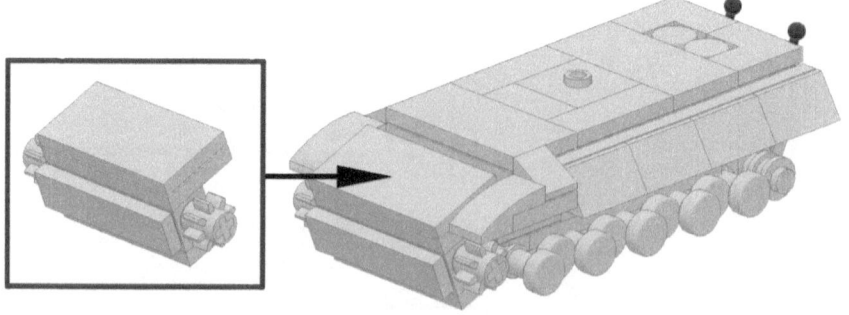

www.ww2custombrickmodels.de

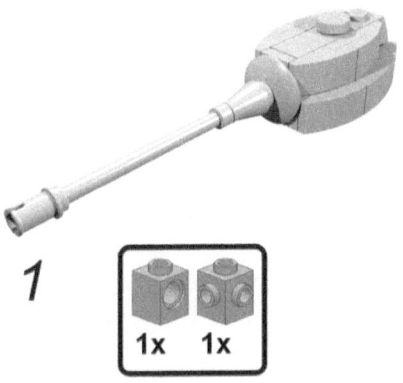

1

2

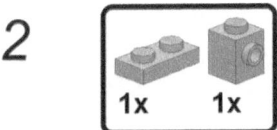

3

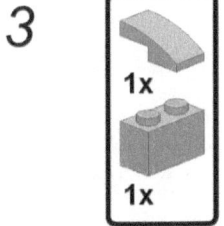

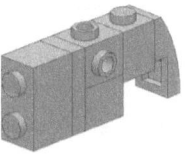

www.ww2custombrickmodels.de

4

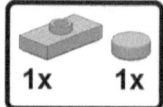

5

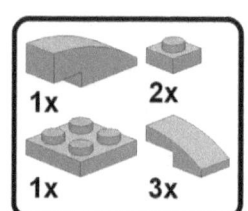

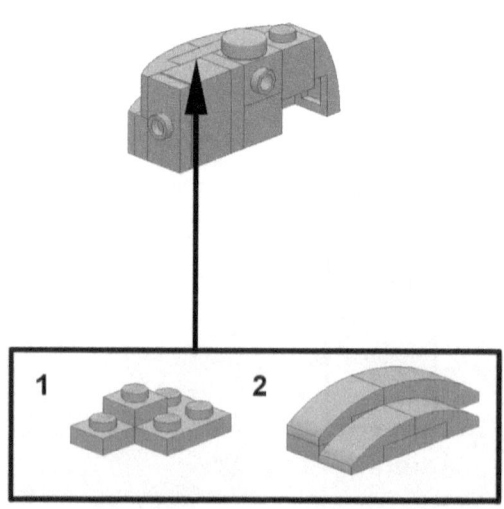

6

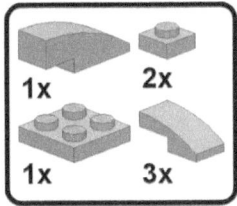

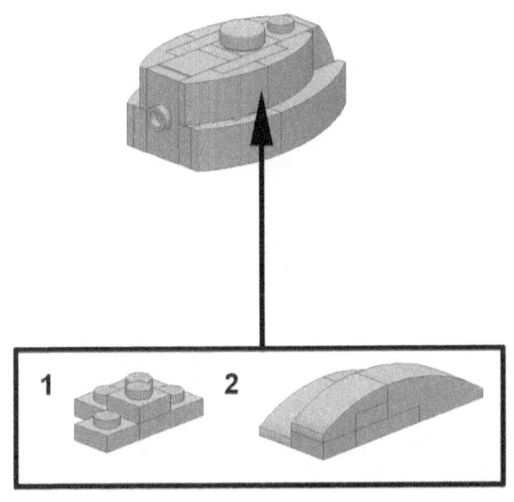

7

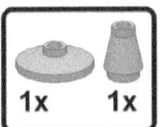

8

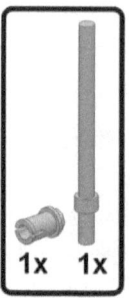

14

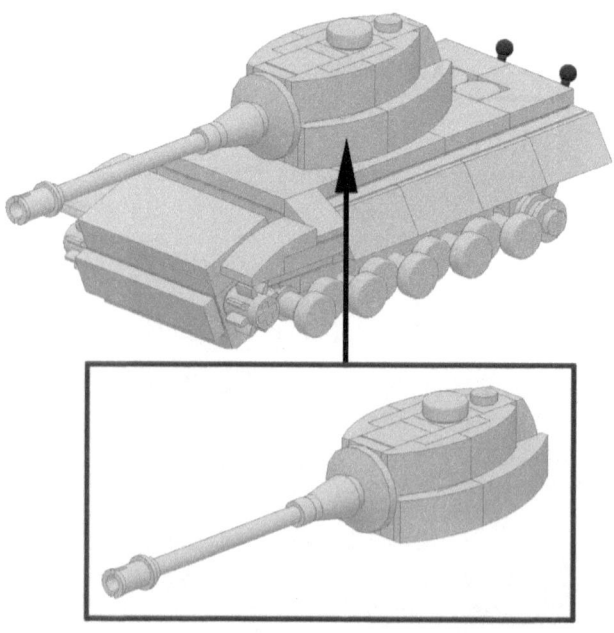

15

 58x

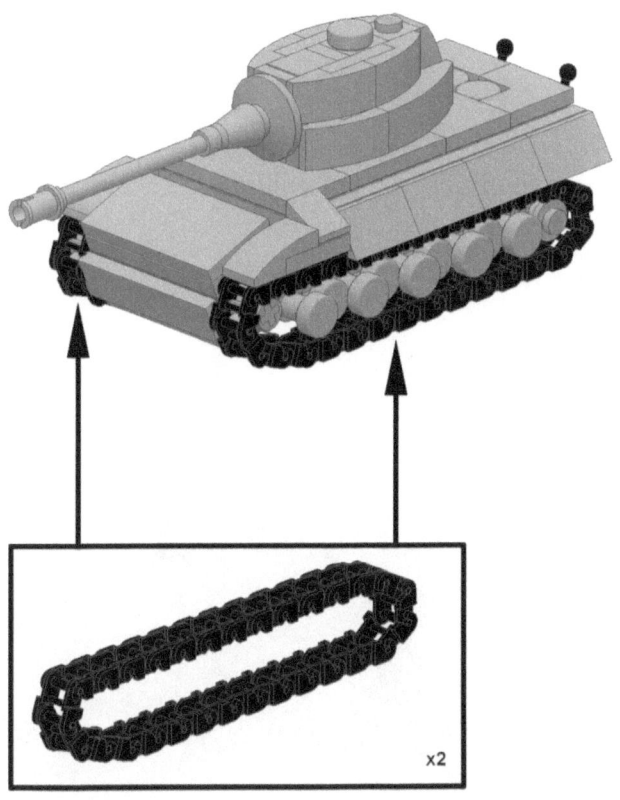

x2

Panzer IV Mini

Zum Bau des Modells benötigen Sie ca. 135 LEGO® Bausteine.
Länge: ca. 8,2 cm Breite: ca. 3,9 cm Höhe: ca. 3,9 cm

Panzer IV Mini

Requires approx. 135 LEGO® bricks.
length: ca. 8.2 cm width: ca. 3.9 cm height: ca. 3.9 cm

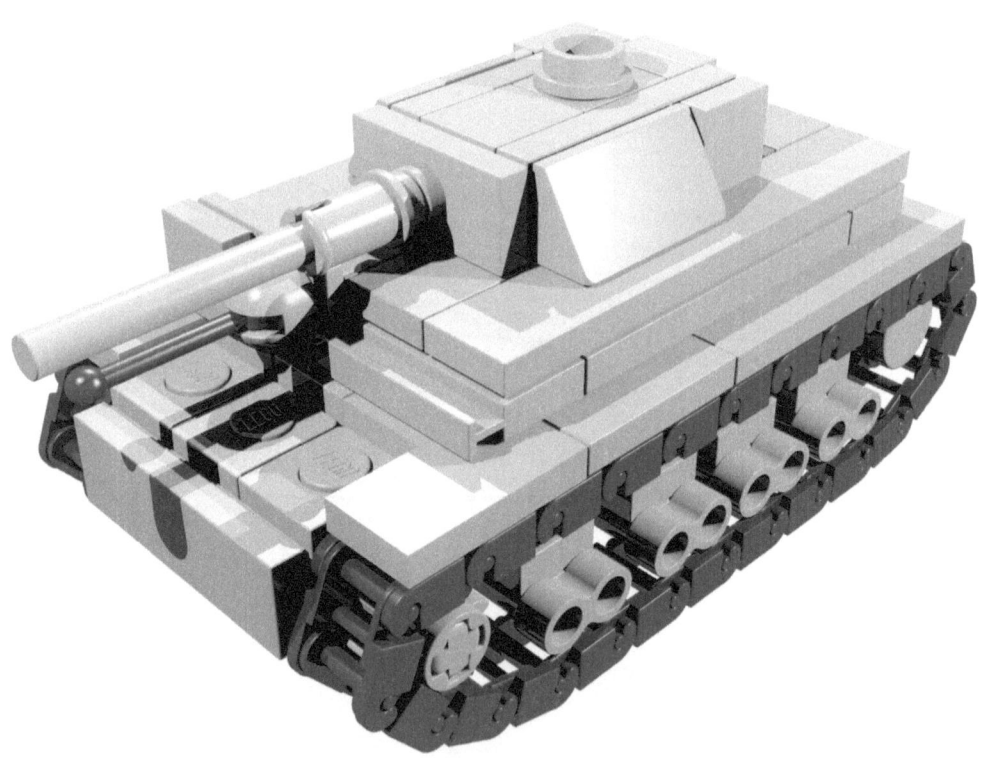

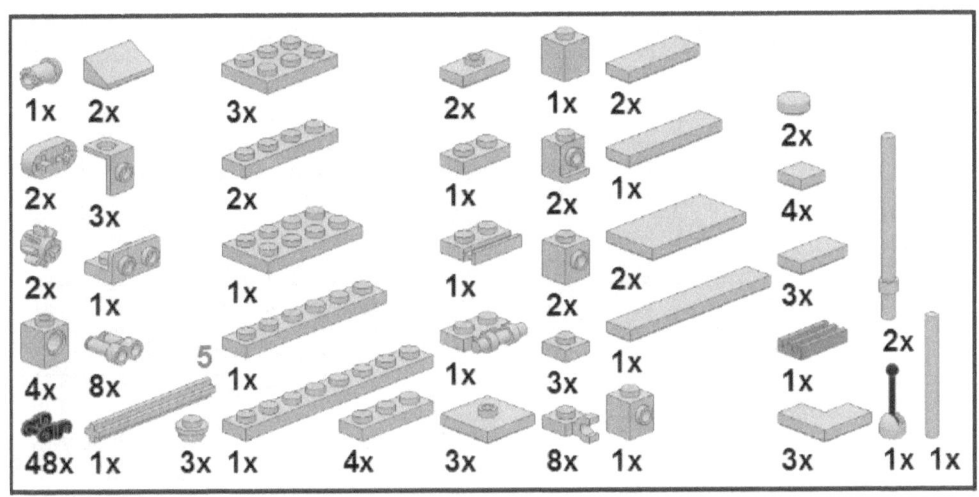

BLink ID	Color	Qty	Description
61252	Grey	8	Plate, Modified 1 x 1 with Clip Horizontal
30162	Grey	8	Minifig, Utensil Binoculars Town
63965	Grey	2	Bar 6L with Stop Ring
41677	Grey	2	Technic, Liftarm 1 x 2 Thin
3460	Grey	1	Plate 1 x 8
87079	Grey	2	Tile 2 x 4
87580	Grey	3	Plate, Modified 2 x 2 with Groove and 1 Stud
6541	Grey	4	Technic, Brick 1 x 1 with Hole
3021	Grey	3	Plate 2 x 3
3666	Grey	1	Plate 1 x 6
2431	Grey	1	Tile 1 x 4
3070b	Grey	4	Tile 1 x 1 with Groove
4070	Grey	2	Brick, Modified 1 x 1 with Headlight
3710	Grey	2	Plate 1 x 4
42446	Grey	3	Minifig, Neck Bracket with Back Stud
63864	Grey	2	Tile 1 x 3
3024	Grey	3	Plate 1 x 1

www.ww2custombrickmodels.de

BLink ID	Color	Qty	Description
3623	Grey	4	Plate 1 x 3
3005	Grey	1	Brick 1 x 1
2540	Grey	1	Plate, Modified 1 x 2 with Handle on Side
3794	Grey	2	Plate, Modified 1 x 2 with 1 Stud (Jumper)
6636	Grey	1	Tile 1 x 6
3020	Grey	1	Plate 2 x 4
32028	Grey	1	Plate, Modified 1 x 2 with Door Rail
3069b	Grey	3	Tile 1 x 2 with Groove
2412b	Dark Grey	1	Tile, Modified 1 x 2 Grille with Bottom Groove / Lip
3023	Grey	1	Plate 1 x 2
14719	Grey	3	Tile 2 x 2 Corner
99780	Grey	1	Bracket 1 x 2 - 1 x 2 Inverted
4592c02	Grey	1	Lever Small Base with Black Lever
87087	Grey	2	Brick, Modified 1 x 1 with Stud on 1 Side
4073	Grey	3	Plate, Round 1 x 1 Straight Side
98138	Grey	2	Tile, Round 1 x 1
3711	Black	48	Technic, Link Chain
47905	Grey	1	Brick, Modified 1 x 1 with Studs on 2 Sides
85984	Grey	2	Slope 30 1 x 2 x 2/3
30374	Grey	1	Bar 4L (Lightsaber Blade / Wand)
4274	Grey	1	Technic, Pin 1/2
3647	Grey	2	Technic, Gear 8 Tooth Type 1
32073	Grey	1	Technic, Axle 5

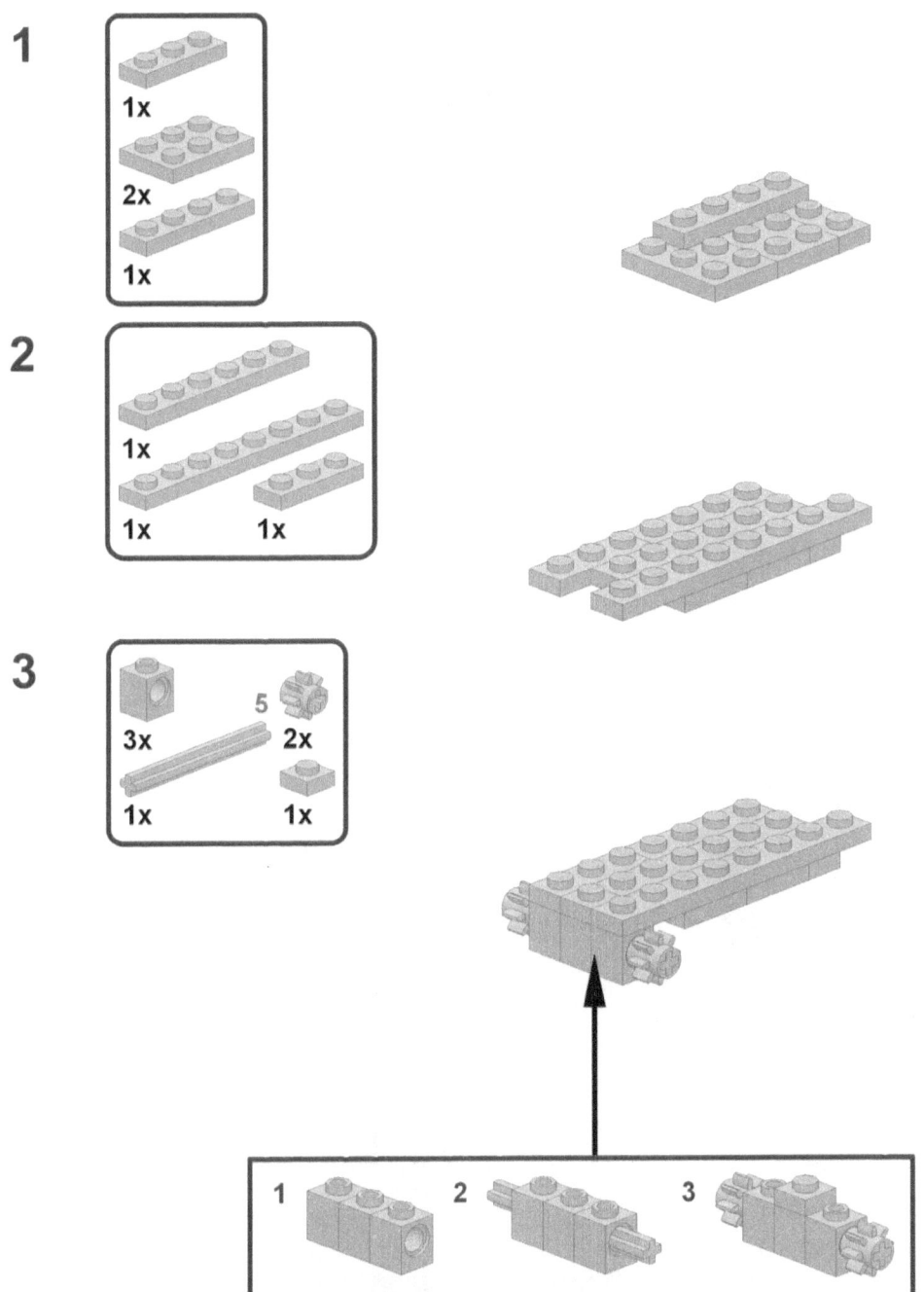

4

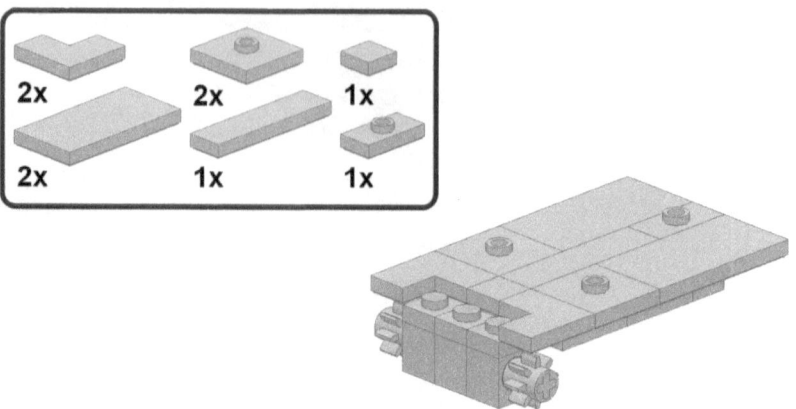

5

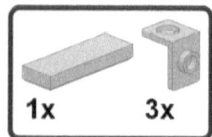

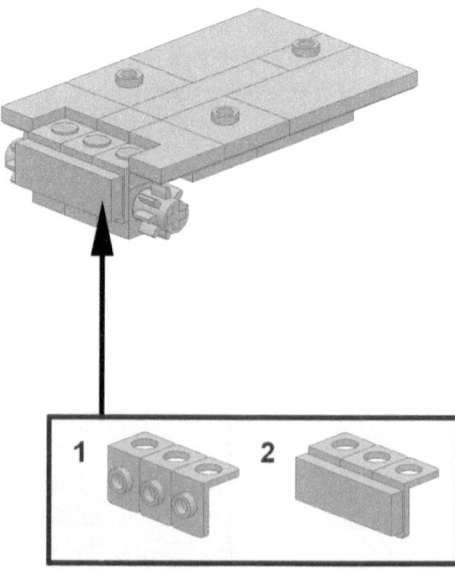

www.ww2custombrickmodels.de

6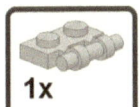

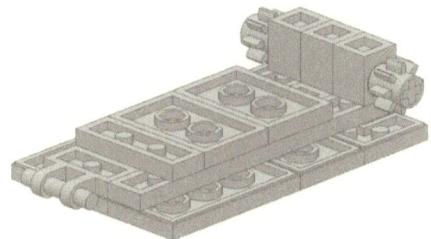

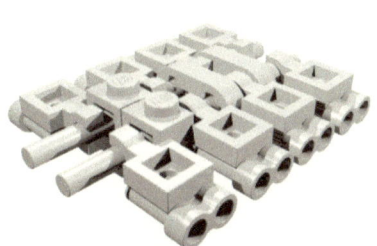

1

2

www.ww2custombrickmodels.de 43

3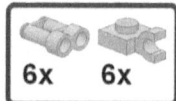

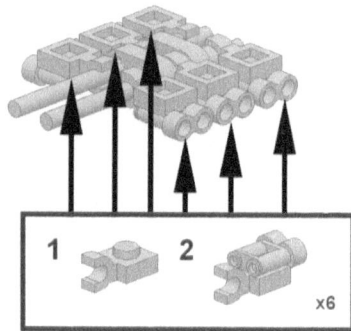

4

5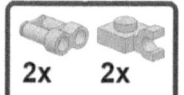

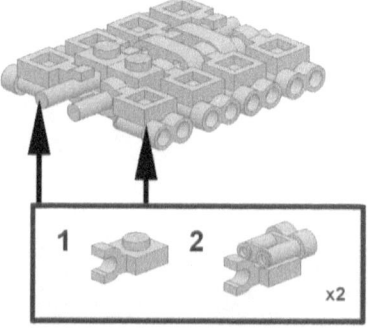

7

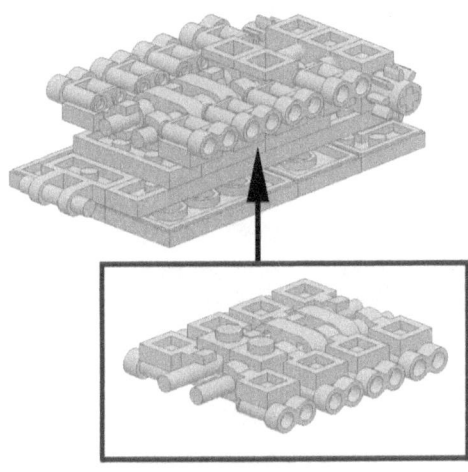

8

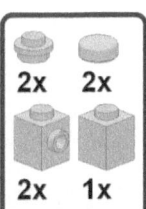

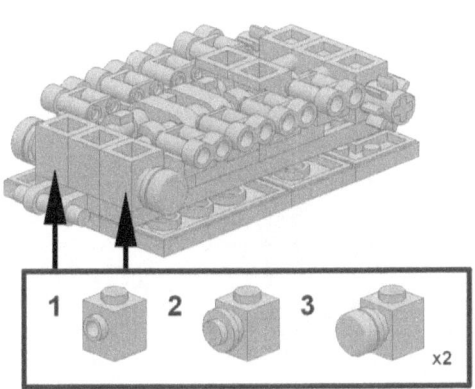

www.ww2custombrickmodels.de   45

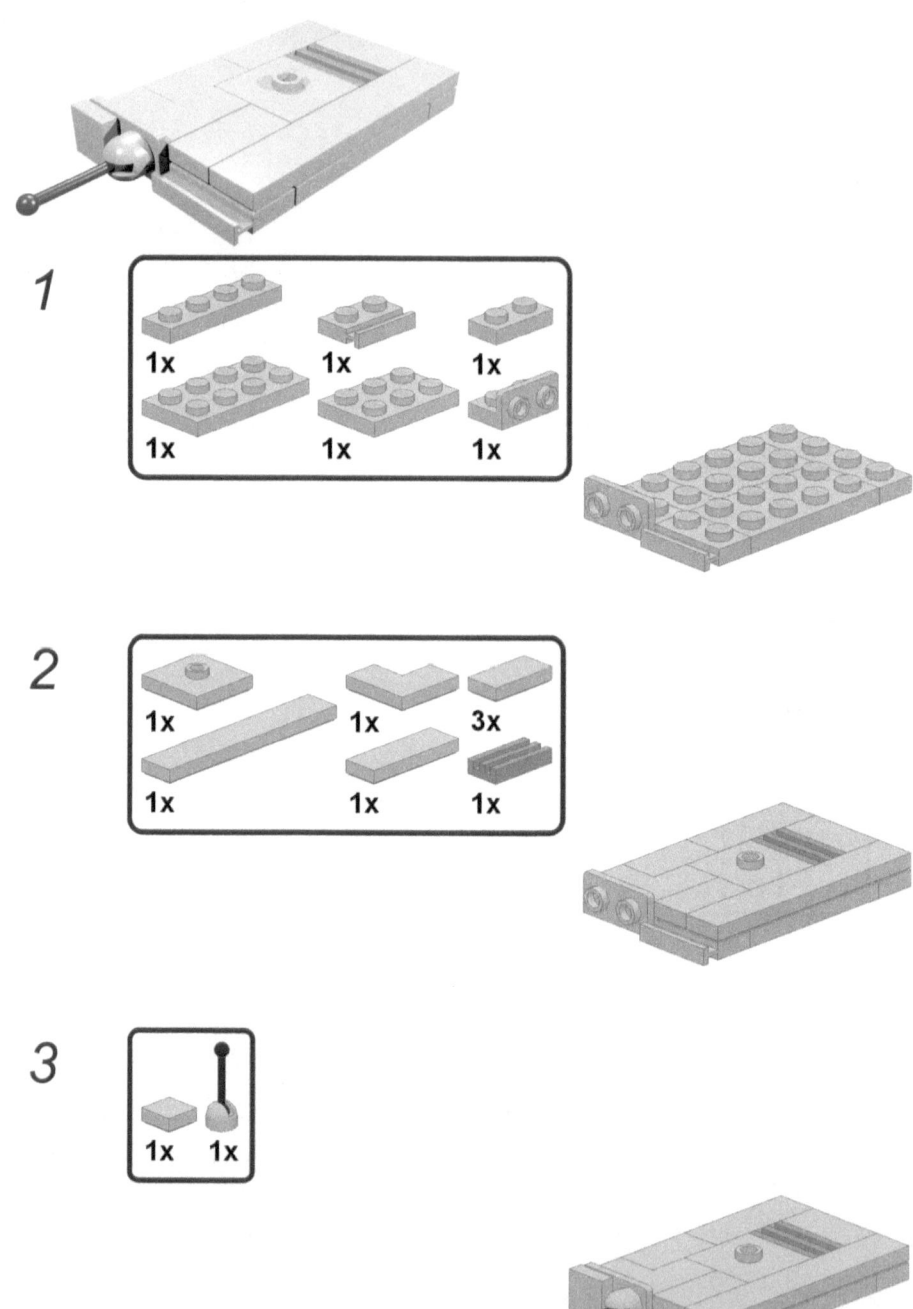

9

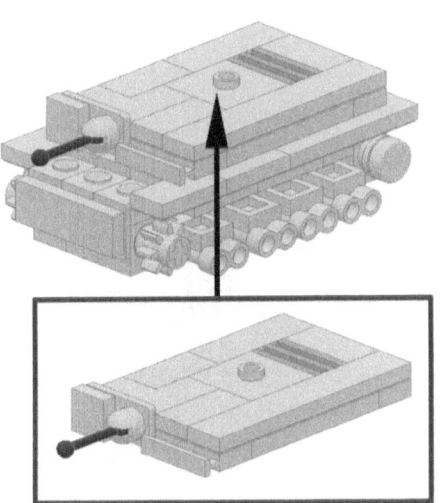

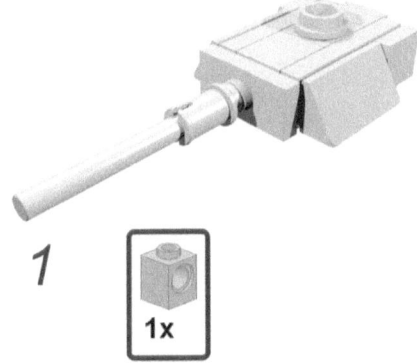

www.ww2custombrickmodels.de 47

2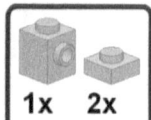

3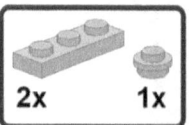

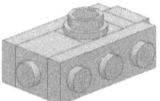

4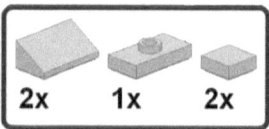

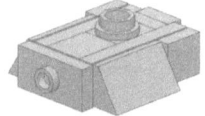

5

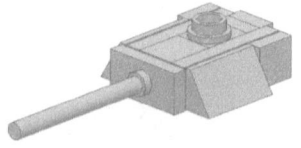

www.ww2custombrickmodels.de

6

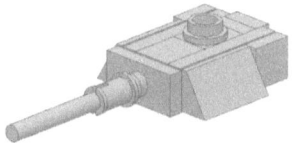

10

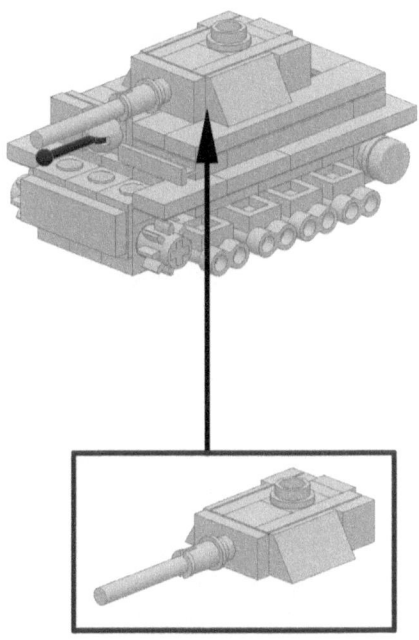

11

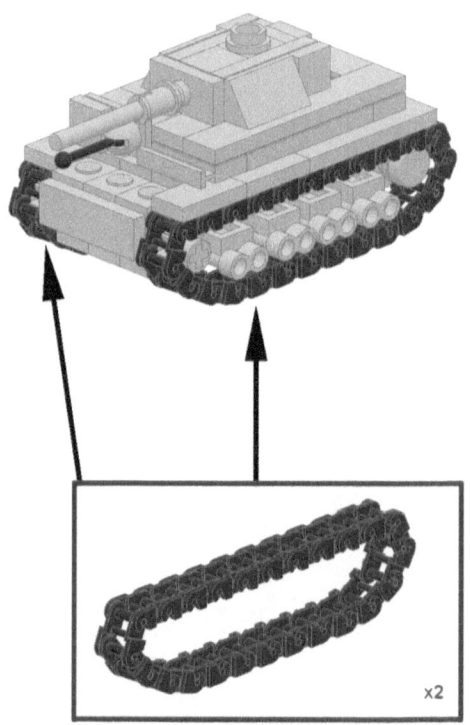

SDKFZ 9 Mini

Zum Bau des Modells benötigen Sie ca. 119 LEGO® Bausteine.
Länge: ca. 10,6 cm Breite: ca. 3,2 cm Höhe: ca. 3,5 cm

SDKFZ 9 Mini

Requires approx. 119 LEGO® bricks.
length: ca. 10.6 cm width: ca. 3.2 cm height: ca. 3.5 cm

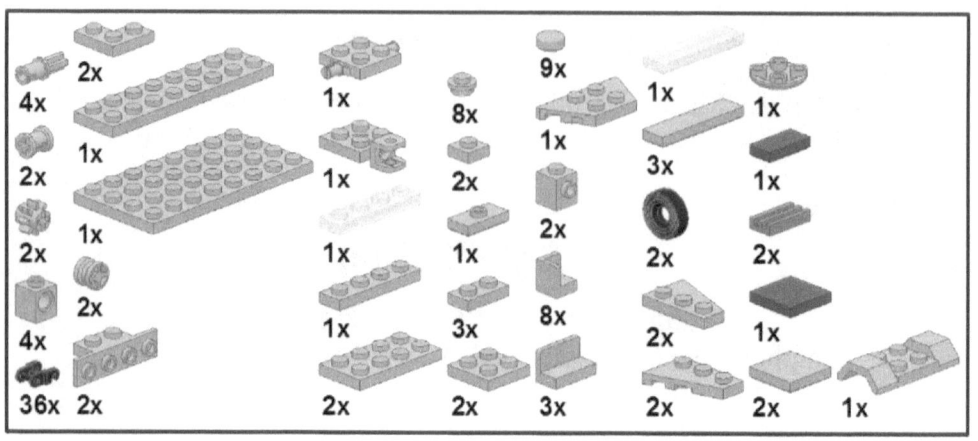

BLink ID	Color	Qty	Description
3711	Black	36	Technic, Link Chain
98138	Grey	9	Tile, Round 1 x 1
3794	Grey	1	Plate, Modified 1 x 2 with 1 Stud (Jumper)
87087	Grey	2	Brick, Modified 1 x 1 with Stud on 1 Side
2412b	Dark Grey	2	Tile, Modified 1 x 2 Grille
2431	Transparent	1	Tile 1 x 4
3710	Transparent	1	Plate 1 x 4
3710	Grey	1	Plate 1 x 4
3068b	Grey	2	Tile 2 x 2 with Groove
43722	Grey	2	Wedge, Plate 3 x 2 Right
43723	Grey	2	Wedge, Plate 3 x 2 Left
51739	Grey	1	Wedge, Plate 2 x 4
2431	Grey	3	Tile 1 x 4
3024	Grey	2	Plate 1 x 1
3023	Grey	3	Plate 1 x 2
3787	Grey	1	Vehicle, Mudguard 2 x 4 with Arch Smooth
3020	Grey	2	Plate 2 x 4
4600	Grey	1	Plate, Modified 2 x 2 with Wheels Holder

BLink ID	Color	Qty	Description
2420	Grey	2	Plate 2 x 2 Corner
4865	Grey	3	Panel 1 x 2 x 1
3035	Grey	1	Plate 4 x 8
6231	Grey	8	Panel 1 x 1 x 1 Corner
3069b	Brown	1	Tile 1 x 2 with Groove
3068b	Brown	1	Tile 2 x 2 with Groove
3022	Grey	2	Plate 2 x 2
3034	Grey	1	Plate 2 x 8
63082	Grey	1	Plate, Modified 2 x 2 with Towball Socket
6541	Grey	4	Technic, Brick 1 x 1 with Hole
2436	Grey	2	Bracket 1 x 2 - 1 x 4
4073	Grey	8	Plate, Round 1 x 1 Straight Side
2654	Grey	1	Plate, Round 2 x 2 with Rounded Bottom
3749	Grey	4	Technic, Axle Pin
3713	Grey	2	Technic Bush
3647	Grey	2	Technic, Gear 8 Tooth Type 1
4624	Grey	2	Wheel 8mm D. x 6mm
3139	Black	2	Tire 14mm D. x 4mm Smooth Small Single

1

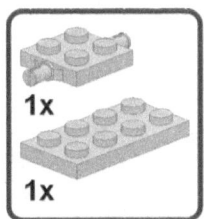

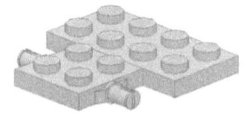

2

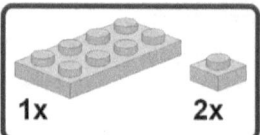

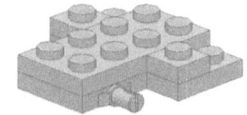

3

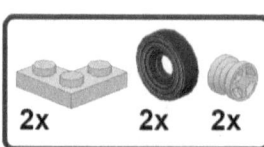

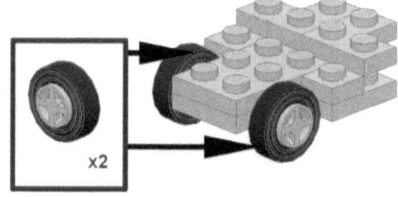

4

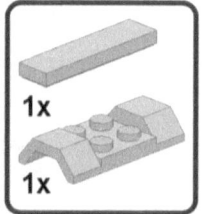

www.ww2custombrickmodels.de

5

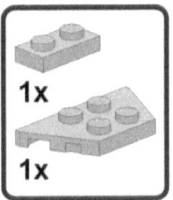

6

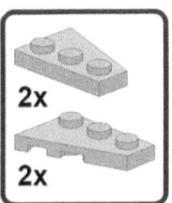

7

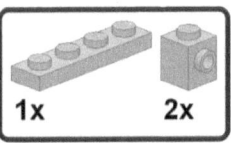

www.ww2custombrickmodels.de

8

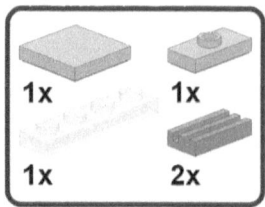 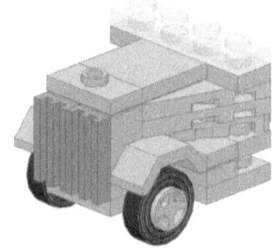

9

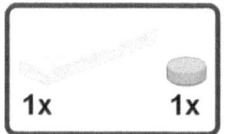 

10

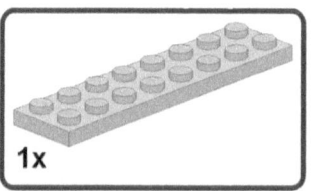 

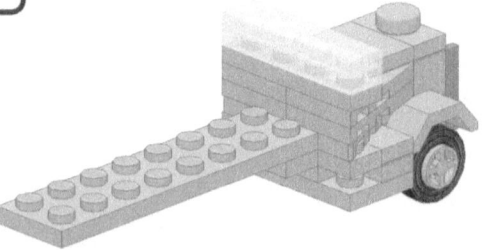

11

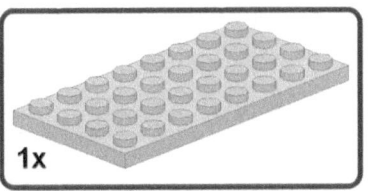

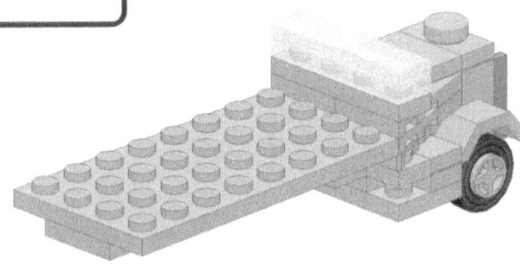

12

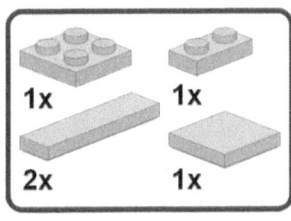

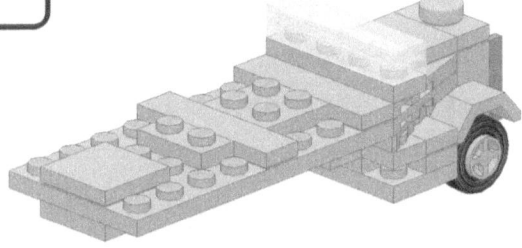

13

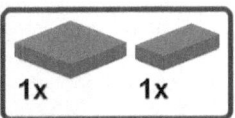

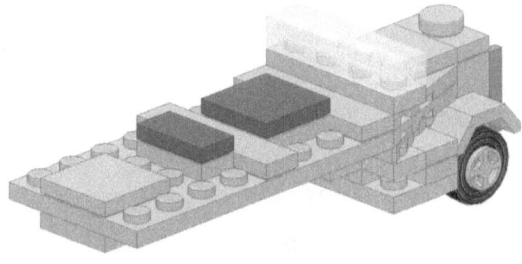

14

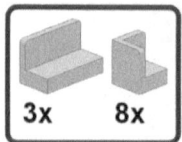

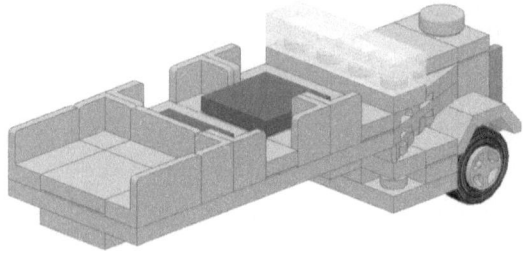

15

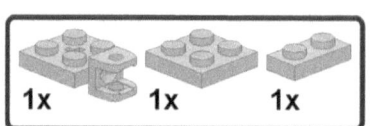

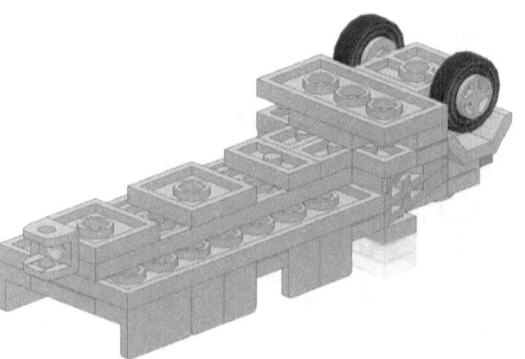

16

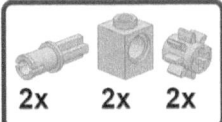

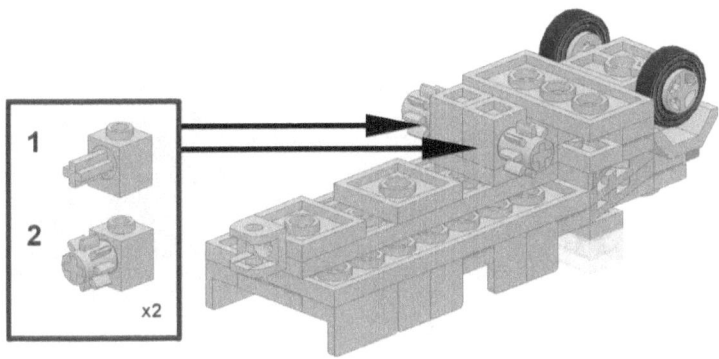

17

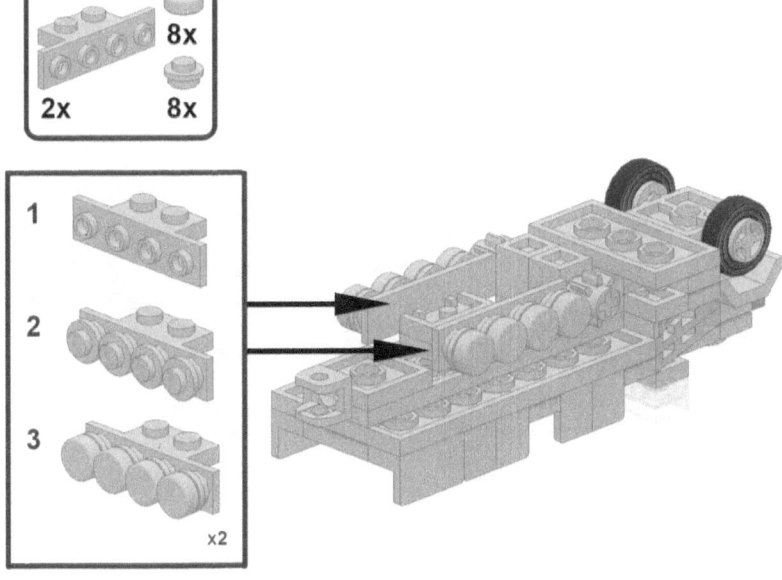

18

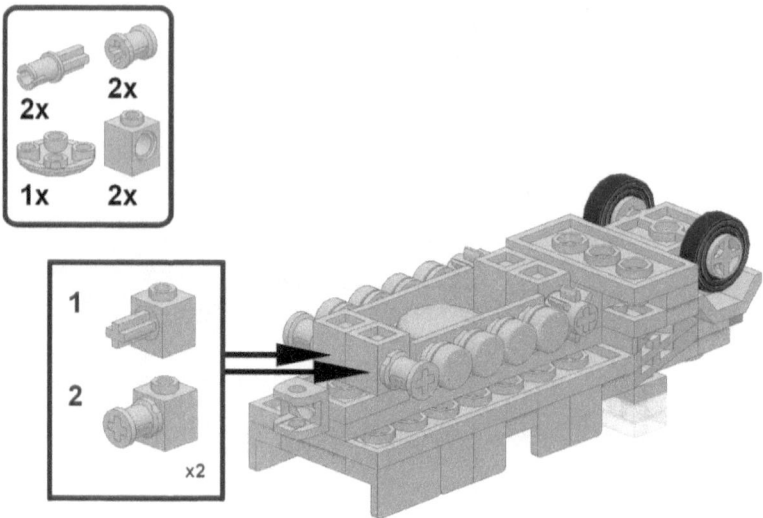

19

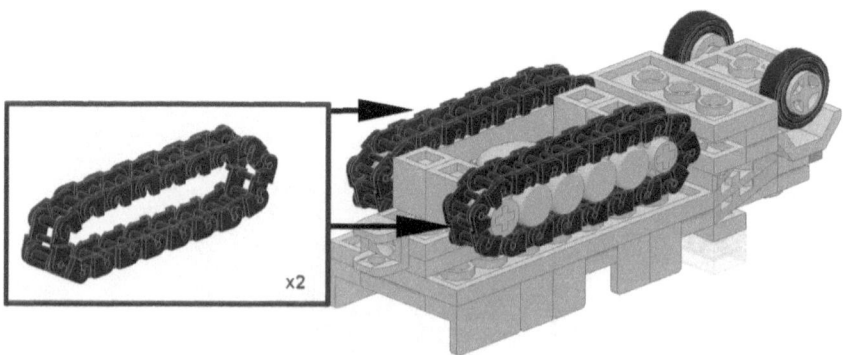

Stug Mini

Zum Bau des Modells benötigen Sie ca. 132 LEGO® Bausteine.
Länge: ca. 9,6 cm Breite: ca. 3,9 cm Höhe: ca. 3,3 cm

Stug Mini

Requires approx. 132 LEGO® bricks.
length: ca. 9.6 cm width: ca. 3.9 cm height: ca. 3.3 cm

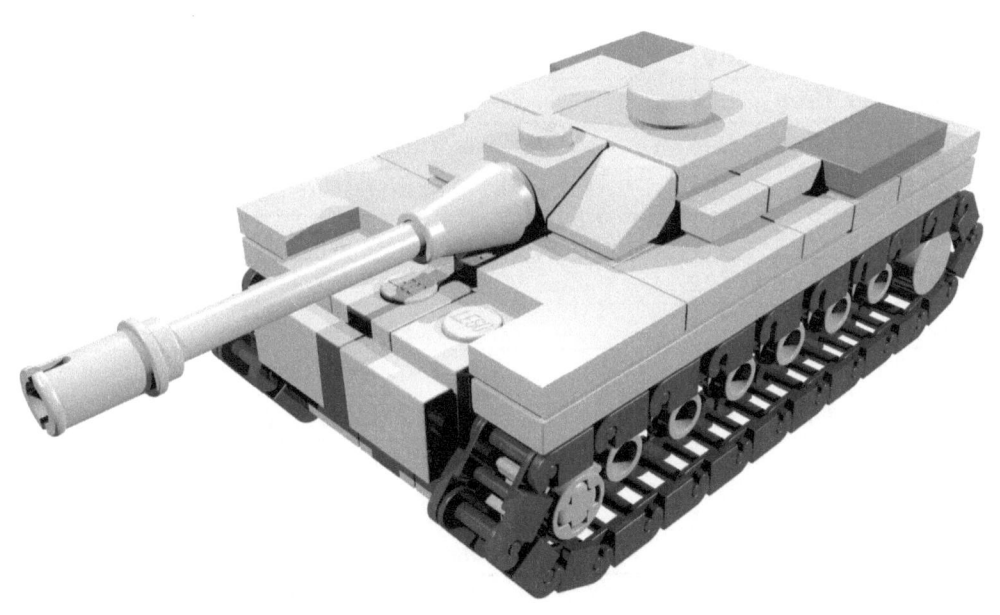

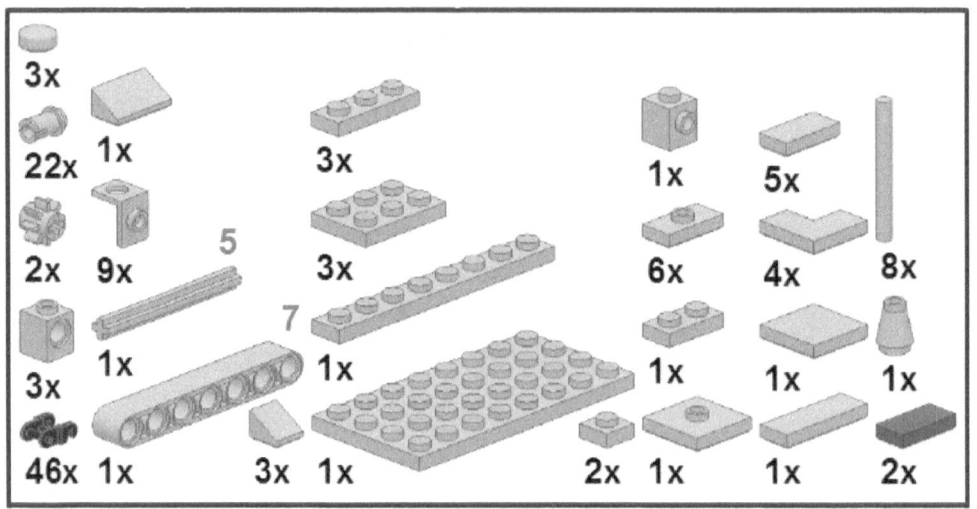

BLink ID	Color	Qty	Description
3021	Grey	3	Plate 2 x 3
42446	Grey	9	Minifig, Neck Bracket with Back Stud
4274	Grey	22	Technic, Pin 1/2
3035	Grey	1	Plate 4 x 8
3024	Grey	2	Plate 1 x 1
63864	Grey	1	Tile 1 x 3
3069b	Grey	5	Tile 1 x 2 with Groove
14719	Grey	4	Tile 2 x 2 Corner
3794	Grey	6	Plate, Modified 1 x 2 with 1 Stud
3623	Grey	3	Plate 1 x 3
54200	Grey	3	Slope 30 1 x 1 x 2/3
85984	Grey	1	Slope 30 1 x 2 x 2/3
6541	Grey	3	Technic, Brick 1 x 1 with Hole
3068b	Grey	1	Tile 2 x 2 with Groove
3460	Grey	1	Plate 1 x 8
32524	Grey	1	Technic, Liftarm 1 x 7 Thick
3023	Grey	1	Plate 1 x 2

BLink ID	Color	Qty	Description
30374	Grey	8	Bar 4L (Lightsaber Blade / Wand)
98138	Grey	3	Tile, Round 1 x 1
32073	Grey	1	Technic, Axle 5
3647	Grey	2	Technic, Gear 8 Tooth Type 1
3069b	Brown	2	Tile 1 x 2 with Groove
4589	Grey	1	Cone 1 x 1 without Top Groove
87087	Grey	1	Brick, Modified 1 x 1 with Stud on 1 Side
87580	Grey	1	Plate, Modified 2 x 2 with Groove
3711	Black	46	Technic, Link Chain

1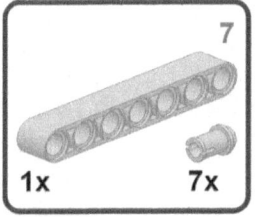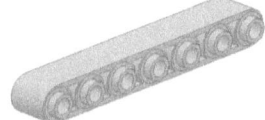

2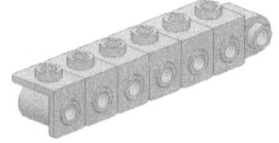

3

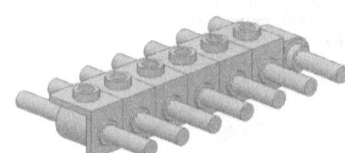

4

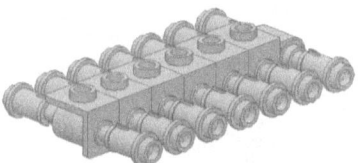

5

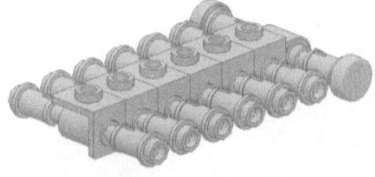

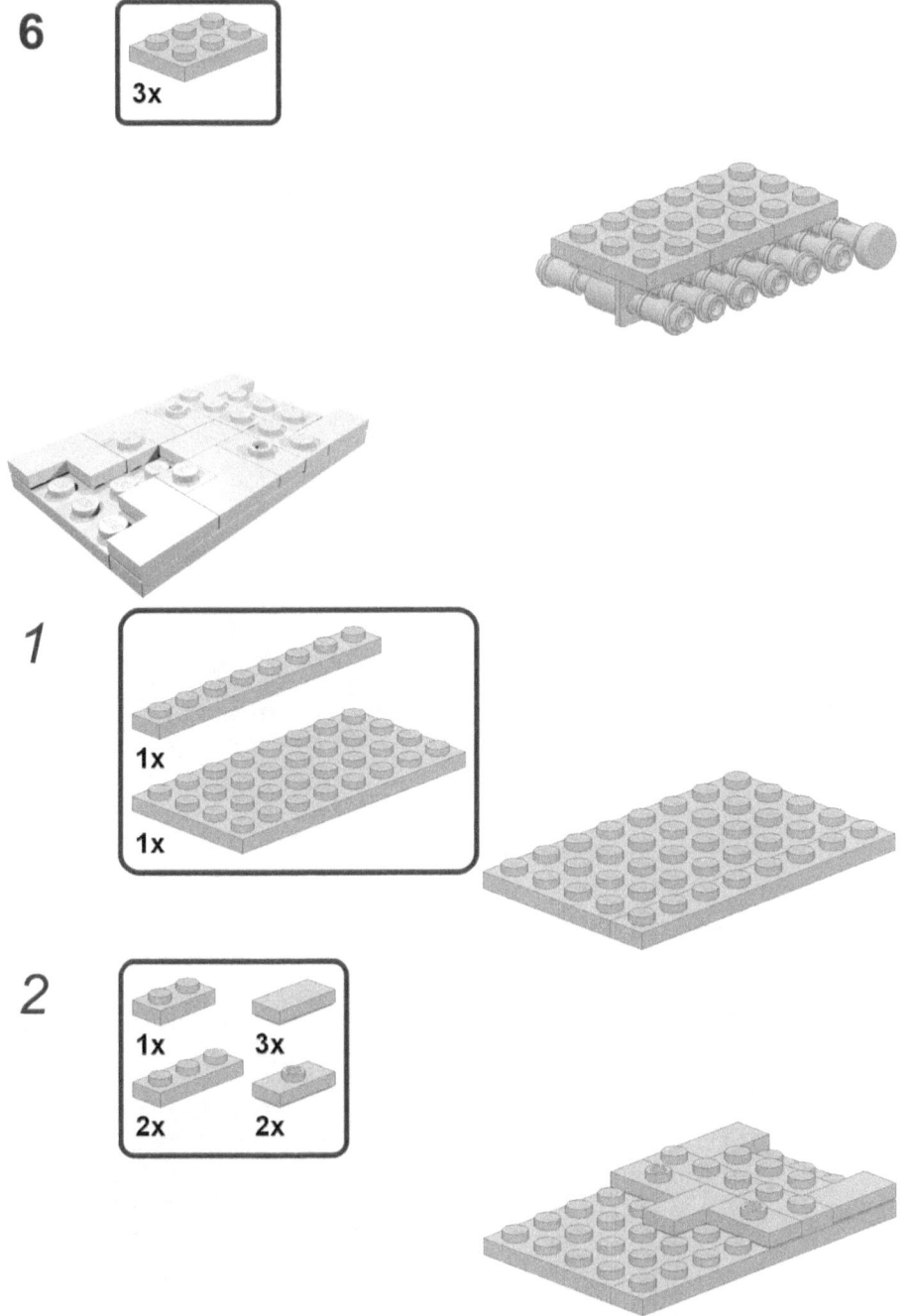

3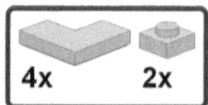

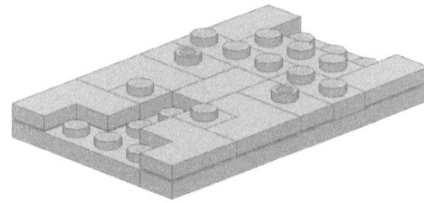

7

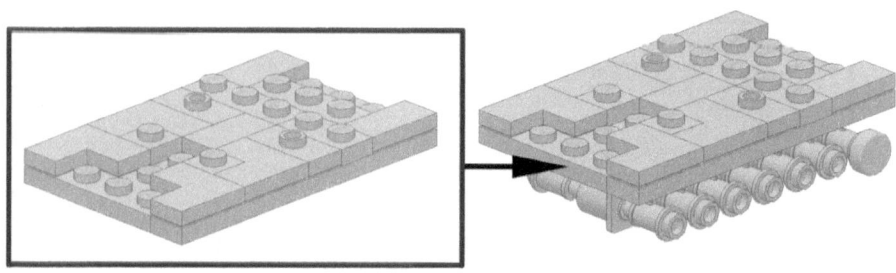

8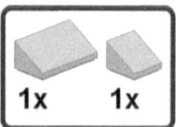

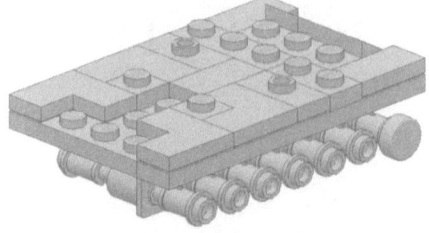

9

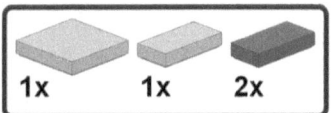

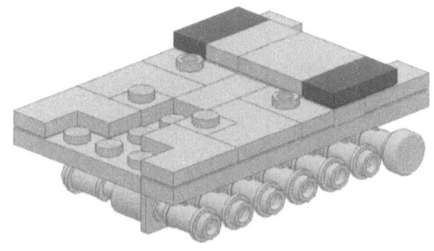

10

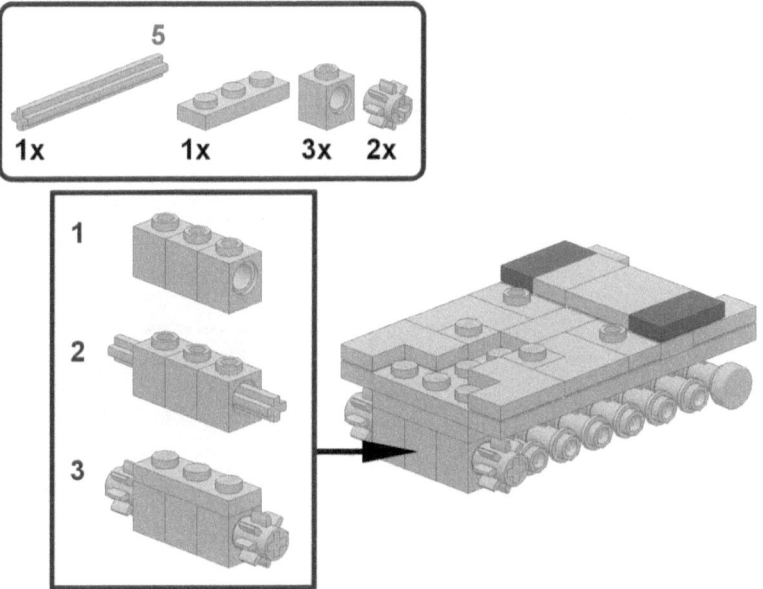

www.ww2custombrickmodels.de 67

11

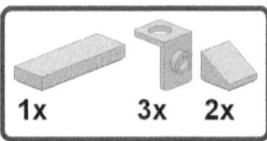

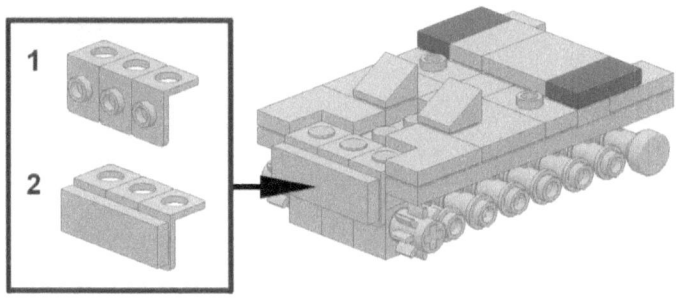

12

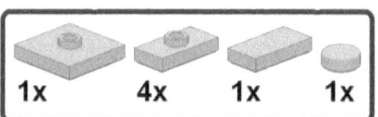

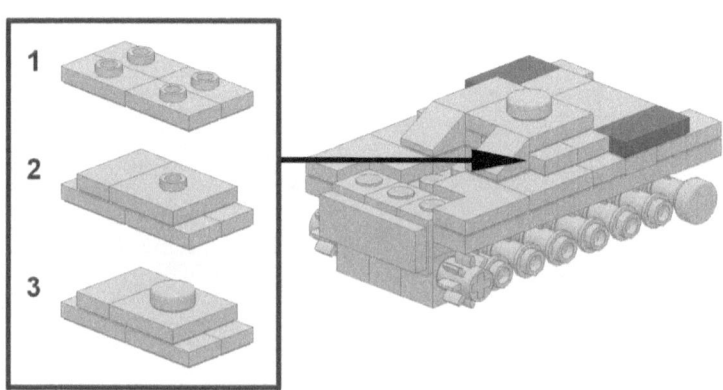

13

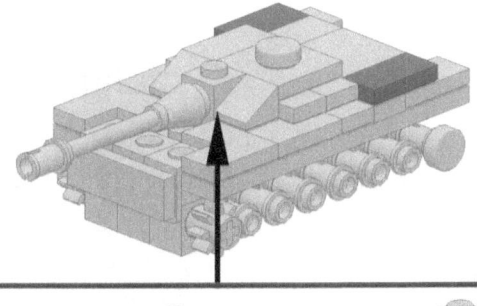

14

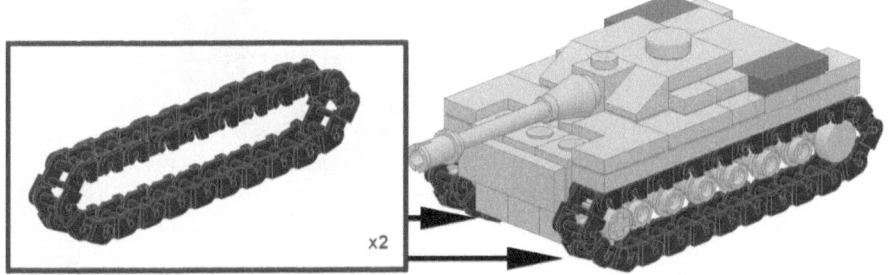

Tiger Mini

Zum Bau des Modells benötigen Sie ca. 192 LEGO® Bausteine.
Länge: ca. 12,5 cm Breite: ca. 5,8 cm Höhe: ca. 4,3 cm

Tiger Mini

Requires approx. 192 LEGO® bricks.
length: ca. 12.5 cm width: ca. 5.8 cm height: ca. 4.3 cm

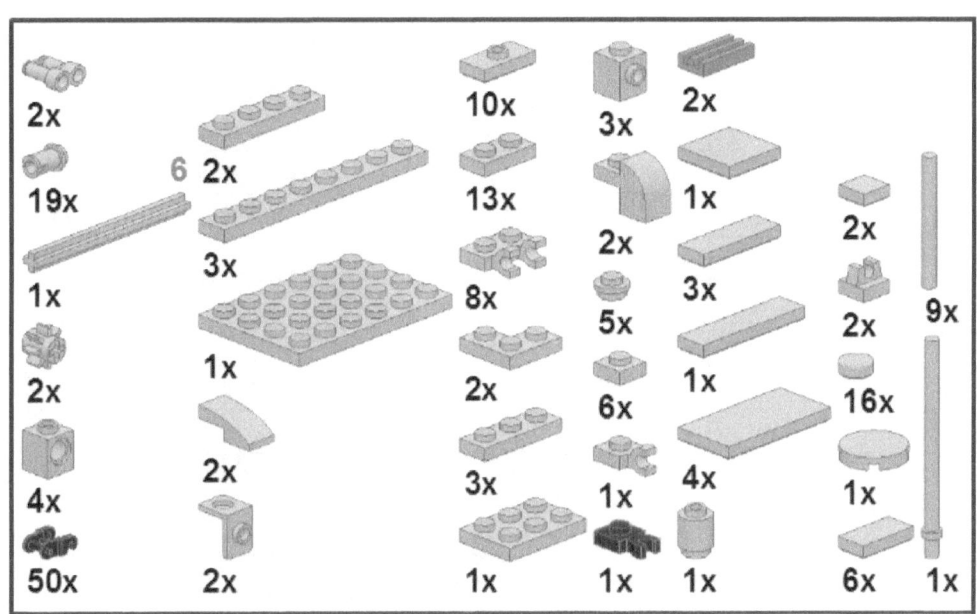

BLink ID	Color	Qty	Description
4073	Grey	5	Plate, Round 1 x 1 Straight Side
4274	Grey	19	Technic, Pin 1/2
30374	Grey	9	Bar 4L (Lightsaber Blade / Wand)
3068b	Grey	1	Tile 2 x 2 with Groove
3023	Grey	13	Plate 1 x 2
60470	Grey	8	Plate, Modified 1 x 2 with Clips Horizontal
3024	Grey	6	Plate 1 x 1
6019	Grey	1	Plate, Modified 1 x 1 with Clip Horizontal
47905	Grey	3	Brick, Modified 1 x 1 with Studs on 2 Sides
15573	Grey	10	Plate, Modified 1 x 2 with 1 Stud
98138	Grey	16	Tile, Round 1 x 1
11477	Grey	2	Slope, Curved 2 x 1 No Studs
42446	Grey	2	Minifig, Neck Bracket with Back Stud
2555	Grey	2	Tile, Modified 1 x 1 with Clip

BLink ID	Color	Qty	Description
3032	Grey	1	Plate 4 x 6
2431	Grey	1	Tile 1 x 4
3069b	Grey	6	Tile 1 x 2 with Groove
2412b	Dark Grey	2	Tile, Modified 1 x 2 Grille with Bottom Groove
63864	Grey	3	Tile 1 x 3
87079	Grey	4	Tile 2 x 4
3460	Grey	3	Plate 1 x 8
4085c	Black	1	Plate, Modified 1 x 1 with Clip Vertical - Type 3
3623	Grey	3	Plate 1 x 3
3062b	Grey	1	Brick, Round 1 x 1 Open Stud
3070b	Grey	2	Tile 1 x 1 with Groove
14769	Grey	1	Tile, Round 2 x 2 with Bottom Stud Holder
3021	Grey	1	Plate 2 x 3
30162	Grey	2	Minifig, Utensil Binoculars Town
3710	Grey	2	Plate 1 x 4
6541	Grey	4	Technic, Brick 1 x 1 with Hole
2420	Grey	2	Plate 2 x 2 Corner
3706	Grey	1	Technic, Axle 6
3647	Grey	2	Technic, Gear 8 Tooth Type 1
4095	Grey	1	Bar 6.6L with Stop Ring (Patio Umbrella Stand)
6091	Grey	2	Brick, Modified 1 x 2 x 1 1/3 with Curved Top
3711	Black	50	Technic, Link Chain

1

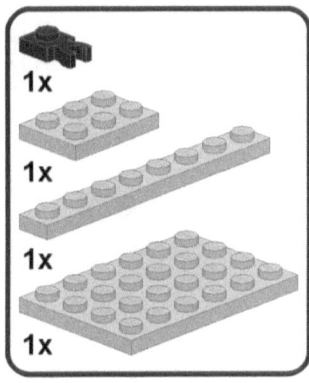
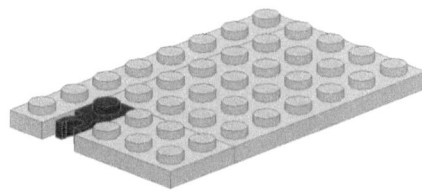

2

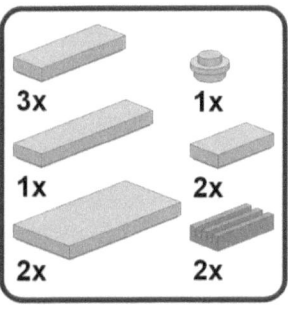
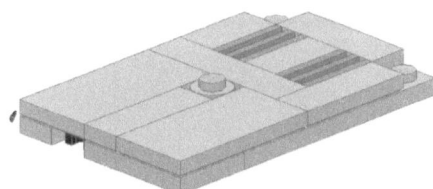

3

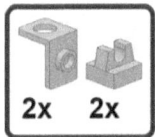
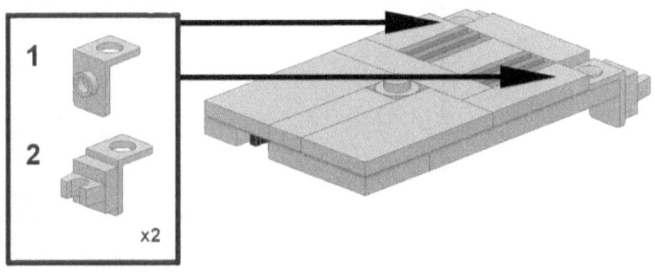

www.ww2custombrickmodels.de

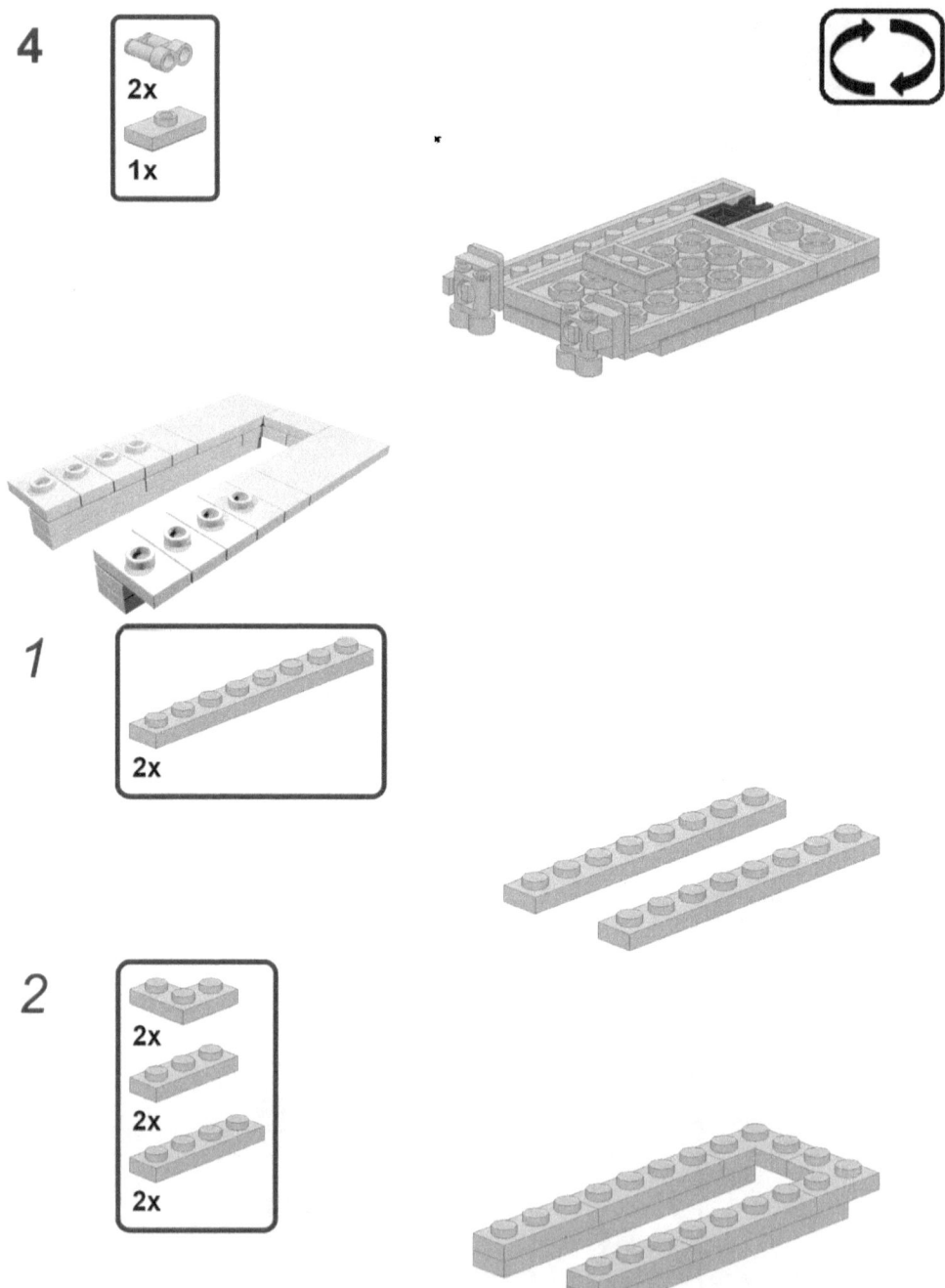

3

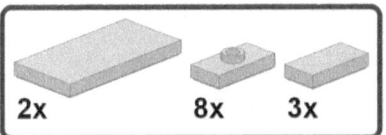

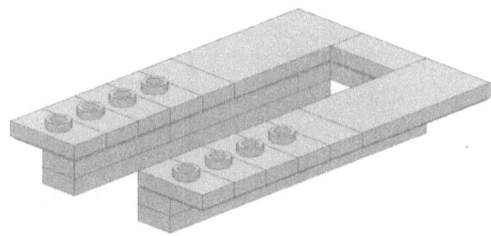

5

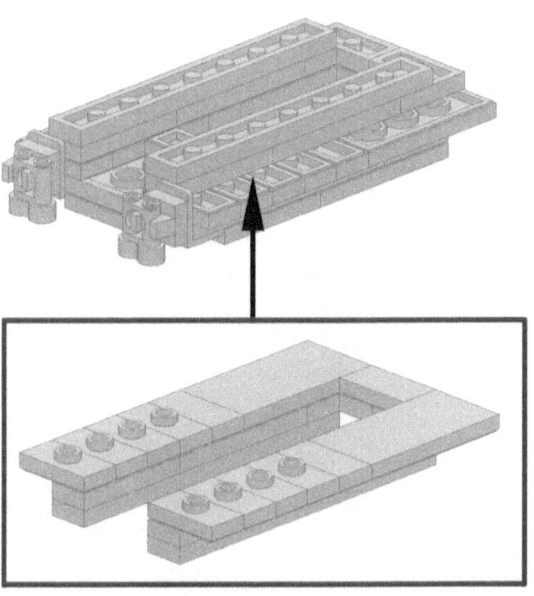

6

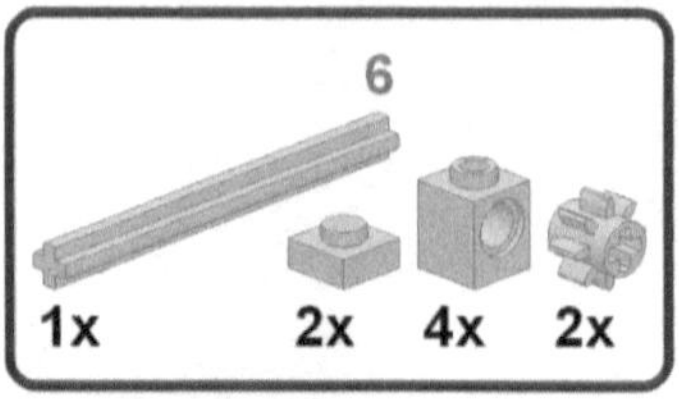

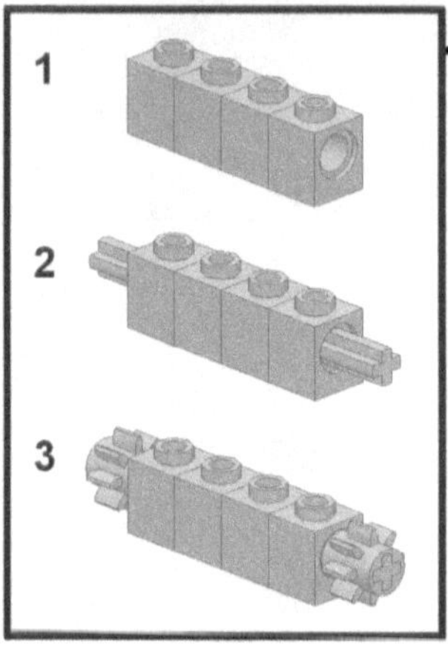

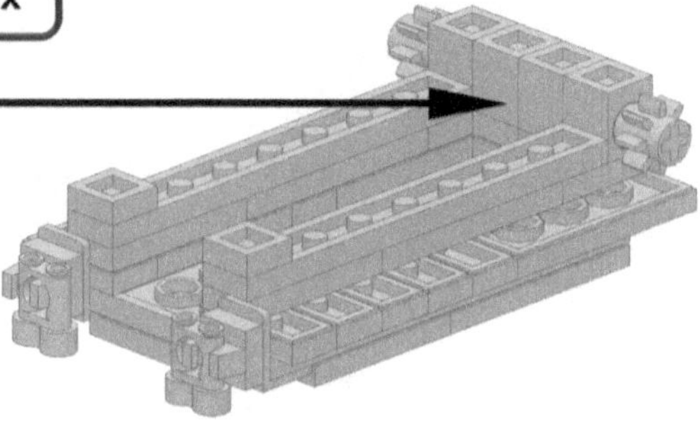

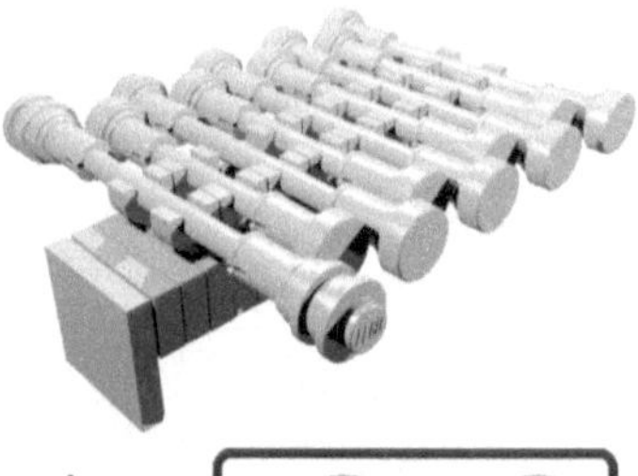

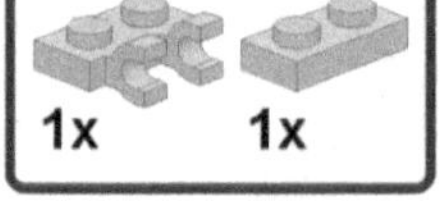

2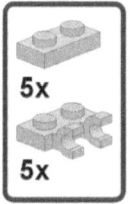

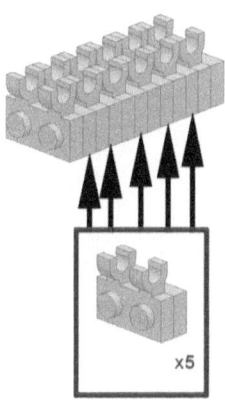

3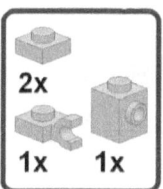

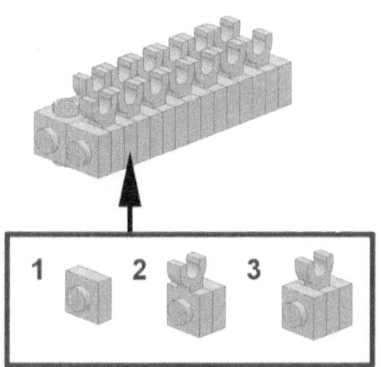

4

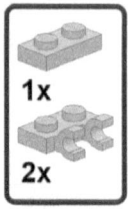

5

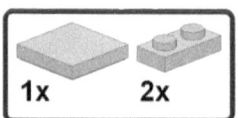
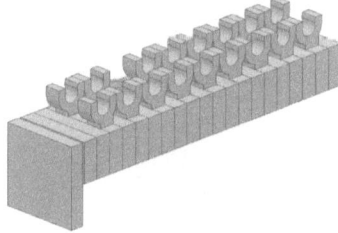

6

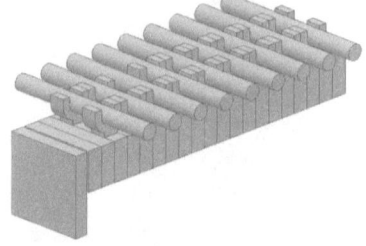

7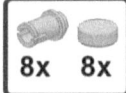

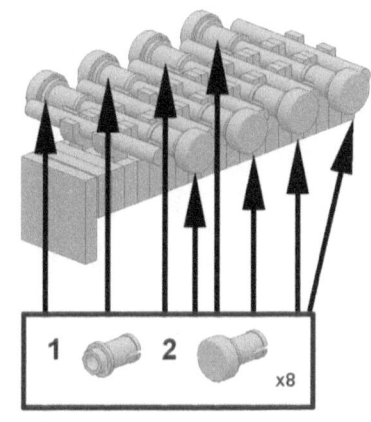

8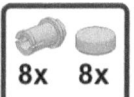

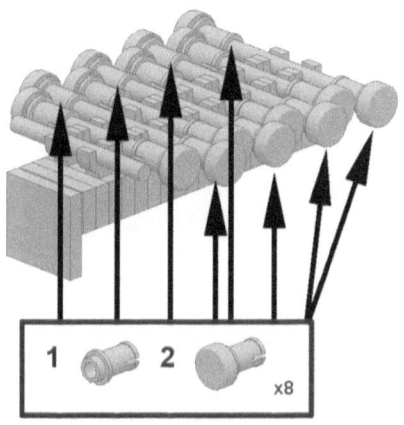

9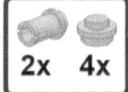

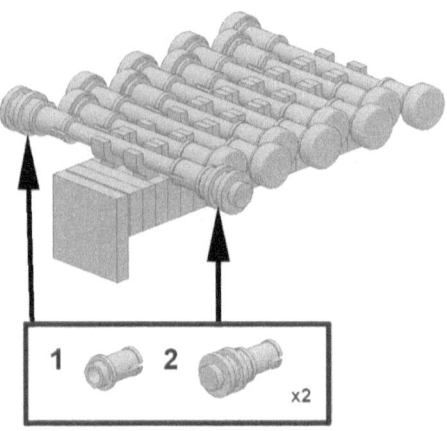

7

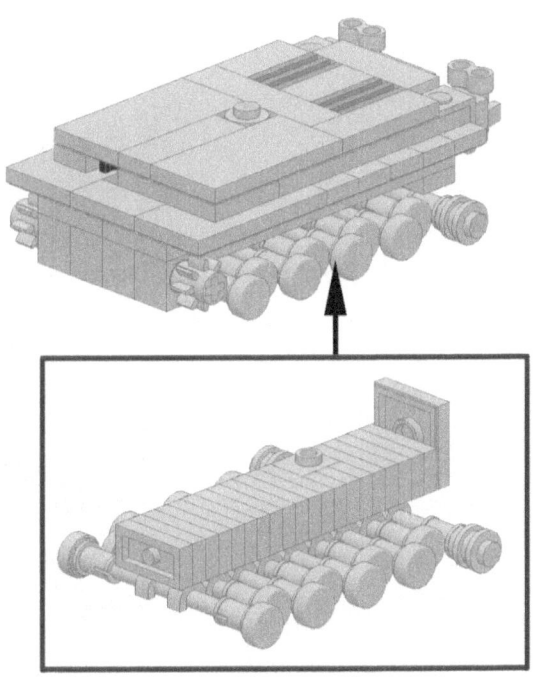

8

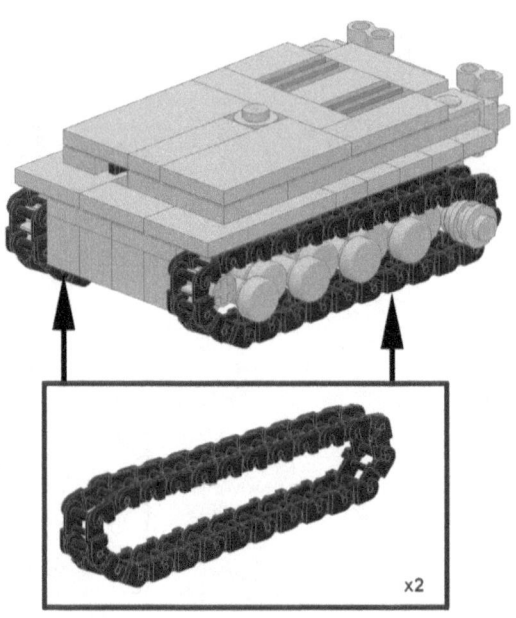

1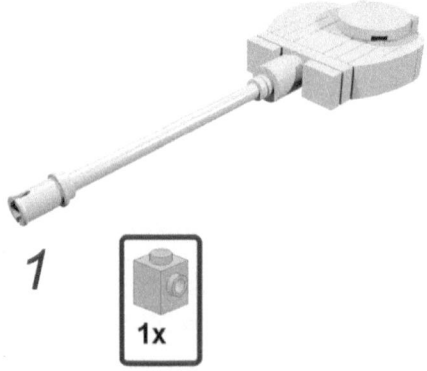

2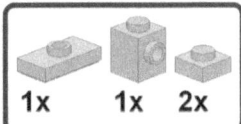

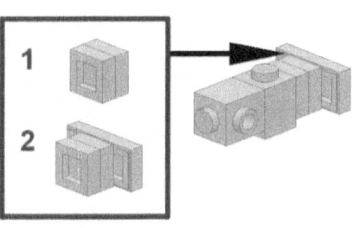

3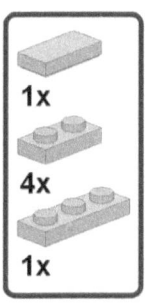

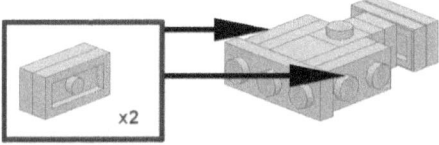

4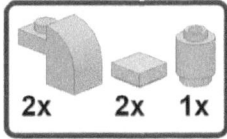

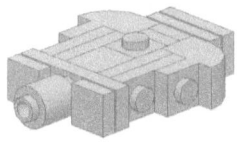

www.ww2custombrickmodels.de

5

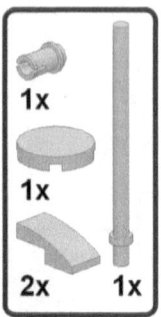

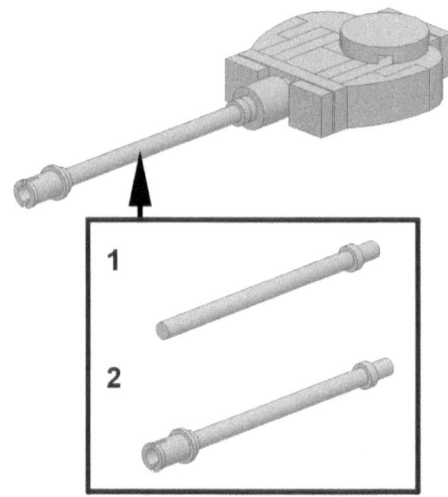

9

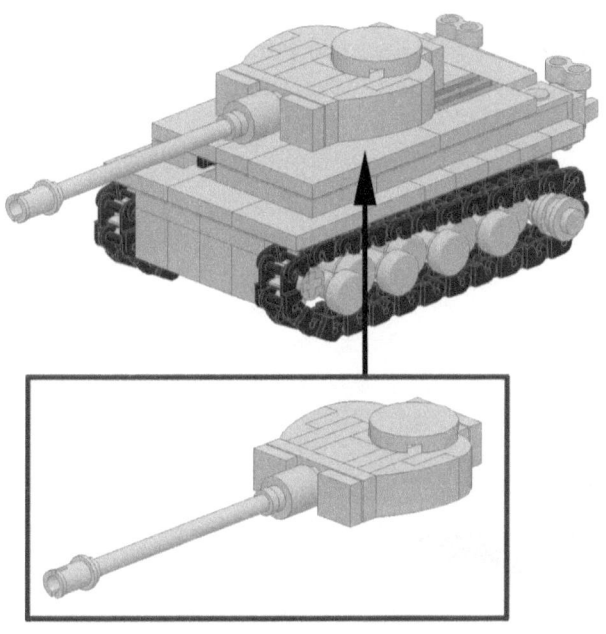

Sie sind auf der Suche nach den Teilen? Besuchen Sie einige der unten stehenden Internetseiten um mit der Hilfe der aufgelisteten Teilenummern dort das passende Teil zu finden.

Not sure where to find the parts? Try some of the websites below. You can search for the parts with the listed part numbers.

www.modernbrickwarfare.com

www.lego.de

www.bricklink.com

www.brickowl.com

Visit this website:

www.brickfactory-berlin.com/

www.ww2custombrickmodels.de

Visit my website:

www.ww2custombrickmodels.de

www.ingramcontent.com/pod-product-compliance
Lightning Source LLC
Chambersburg PA
CBHW030446220526
45464CB00006B/2436